How to Classify Your Medical Device Under European Regulations

How to Classify Your Medical Device Under European Regulations

Shalinee Naidoo

ARCLER PRESS

www.arclerpress.com

How to Classify Your Medical Device Under European Regulations

Shalinee Naidoo

Arcler Press
224 Shoreacres Road
Burlington, ON L7L 2H2
Canada
www.arclerpress.com
Email: orders@arclereducation.com

ISBN: 978-1-77469-187-8 (Hardcover)

ABOUT THE AUTHOR

Shalinee Naidoo is currently the Regulatory and Product Development Manager of a medical device manufacturer based in South Africa. She is directly involved in global regulatory compliance and the design and development of new medical devices from idea conception to market. She is also the founder of Life of Shal (www.lifeofshal.com) an online travel journal created to inspire others to explore the world and Scientist's Sanctuary (www.scientistssanctuary.com) – a science communication company that specializes in bridging the gap between scientific knowledge and creative communication for both the academic and corporate world.

TABLE OF CONTENTS

LIST OF FIGURES

LIST OF TABLES

LIST OF BOX

LIST OF ABBREVIATIONS

AEDs	Automated External Defibrillators
AHWP	Asian Harmonization Working Party
AIMD	Active Implantable Medical Devices
AIMDD	Active Implantable Medical Device Directive
AR	Authorized Representatives
ASR	Articular Surface Replacement
BSI	British Standards Institution
CA	Competent Authorities
CNMD	Classification Names for Medical Devices
ECRI	Emergency Care Research Institute
EDMA	European Diagnostic Manufacturers Association
EEA	European Economic Area
ELISA	Enzyme-Linked Immunosorbent Assay
ER	Essential Requirements
EUDAMED	European Database for Medical Devices
FDA	Food and Drug Administration
GHTF	Global Harmonization Task Force
GMDN	Global Medical Device Nomenclature
GMP	Good Manufacturing Practices
HF	High Frequency
HPRA	Health Products Regulatory Authority
HRTC	Health Registration and Training Center
ICIJ	International Consortium of Investigative Journalists
IEC	International Electrotechnical Commission
IFU	Instructions for Use
IMDRF	International Medical Device Regulators Forum
ISO	International Organization for Standardization

IUDs	Intrauterine Devices
IVD	*In Vitro* Diagnostic
IVDD	*In Vitro* Diagnostic Devices Directive
IVDMDD	*In Vitro* Diagnostic Medical Device Directive
IVDR	*In-Vitro* Diagnostic Regulation
MD	Medical Device
MDA	Medical Device Alert
MDD	Medical Device Directive
MDR	Medical Devices Regulation
MHRA	Medicines and Healthcare Products Regulatory Agency
MoM	Metal-on-Metal
MRI	Magnetic Resonance Imaging
NB	Notified Bodies
NHS	National Health Service
NJR	National Joint Registry
NKKN	Norsk Klassifisering Koding and Nomenklatur
PAHO	Pan American Health Organization
PIP	Poly Implant Prothese
QMS	Quality Management System
SME	Small to Medical Enterprise
UDI	Unique Device Identifier
UMDNS	Universal Medical Device Nomenclature System
WHO	World Health Organization

INTRODUCTION

Medical devices are essential in the delivery of quality health care. They enable health care providers to diagnose, treat, monitor, and provide therapy to patients all around the world. Without properly regulated medical devices, routine medical procedures, whether they be bandaging an open wound or implanting an artificial hip, would be impossible. While medical devices continue to be regulated, advancing technology is producing an overwhelming number of new devices onto the market. One of the biggest issues affecting the medical device industry is the diversity of devices available, which often affects the overall degree of complexity, applications of use, users, and related issues. As a result, medical devices will be most effective when considered in the wider context of the health industry, taking all the above-mentioned parameters into consideration together with the main aim of addressing what the overall public health needs are. These often include prevention, clinical care, investigation, diagnosis, treatment, follow-up, and overall access to appropriate health.

Quality management of medical devices helps ensure that these devices and related services are provided in a safe and effective way. If you are a manufacturer who wants to place their medical device on the European market, you will need to make sure it complies with the specific European Directives set forth by the European Commission. In this case it will be the medical device directives (MDD): AIMDD 90/385/EEC; MDD 93/42/EEC; IVDMDD 98/79/EC and more recently the Medical Device Regulation 2017/745. This means you will need to provide your device complies with the essential requirements (ER) of these Directives/Regulations before you can affix a CE mark onto it. For that purpose, the device must go through the CE marking process. The direction of the process will depend on the actual class of the medical device and related choice of conformity assessment route. Often this can become confusing to all parties involved in the details around how the device was designed as well as the overall intended use is not clearly defined. This volume aims to provide an easy-to-understand guide for beginners to the medical device regulations in Europe with a specific focus on classification methods. It looks specifically at how to class a medical device based on the risk associated with it as well the details around the European Classification Systems provided in the MEDDEV 2.4/1. This volume also delves into the detail around defining borderline medical devices and how they are classified according to the Manual on Borderline and Classification in the Community Regulatory Framework for Medical Devices as published by the European Commission.

PREFACE

Over the last 20 or so years, the number, range, and complexity of medical devices available on the market has increased drastically and as a result, so has the complexity of the regulations involved. Each device brings with it its own set of risks and this in itself makes the opportunity for misuse considerable. With new and emerging technologies as well as various well-known incidents within the medical device industry, the current regulatory framework has since been challenged. In fact, many gaps and scarcity of skills and expertise have been identified. For this reason, there was an increasing need to update the current medical device directive (MDD 93/42/EEC) in the European Union, which in turn led to the development and release of the Medical Device Regulation (EU MDR 2017/745). The release of the new Medical Device Regulation (EU MDR 2017/745) in the Official Journal of the European Union in May of 2017 marked the start of a 3-year transition period for manufacturers, suppliers, notified bodies (NB), and National Competent Authorities (CA). Compliance to the new regulation was set to take effect in May of 2020 but has now been postponed by a year due to the coronavirus pandemic. Complying to either current or new regulation requires a manufacturer to thoroughly understand the intended use of their device as well as the risks associated with it before they can begin classify their device and determining the best route to show compliance. This volume provides an overview of how to classify a medical device under the European Regulations with particular focus on the related guidance documents.

Chapter 1 Introduction To Medical Devices

CONTENTS

1.1. EFFECT OF GLOBAL TRENDS ON MEDICAL DEVICE DEMANDS

In order to understand and predict global trends on overall medical device demands, research must target and consider regional trends and differences in the overall proportional burden of diseases and disabilities. It is a well-known fact that globally, life expectancy is increasing, and as such, more and more people are living well into their 80s and 90s. These aging populations seem to be growing more quickly in less developed countries; with the majority of older populations living predominately in rural areas where Asia and Africa are concerned.

Longer life expectancy is often associated with the presence of several concurrent illnesses (also known as co-morbidity) and chronic debilitating conditions such as coronary heart disease, cancer, and dementia. Aging populations coupled with decreasing fertility and overall lower birth rates is estimated to lower population size and reduce the labor force available. In addition, old-age dependency is expected to increase as well as the average frequency of ill-health and disabilities experienced. Collectively, all these factors are thought to raise the need for medical devices as well as the training of patients, users, and caregivers that will be using these devices. It should also be noted that in addition to non-communicable diseases, many developing countries are also associated with diseases brought on by poverty such as various communicable diseases, nutrient-related conditions, malaria, and various parasitic diseases; all of which affect the overall demand for medical devices.

1.2. URBANIZATION

Urbanization is one of the major factors affecting the overall demand for medical devices. Urban growth rates are estimated to be the highest in developing countries, with expected increases in the rural areas of Asia and Africa over the next few years. It is estimated that by 2050, the urban population of developing countries will reach approximately 5.3 billion people. The urban population of developed and/or industrialized countries are expected to rise ever so slightly from 900 million in 2005 to 1.1 billion by 2050. It is also estimated that most urban residents will be poorer, and access to suitable health care will become more and more difficult. This is expected to subsequently heighten the need for primary health care and suitable medical devices to treat varying conditions.

1.3. REDUCED WORKERS

In order to ensure a properly functioning health care system, countries must ensure they have a healthy and large workforce that have a targeted list of strategies to work through. The World Health Organization (WHO) estimates that currently, there is a global shortage of doctors, nurses, midwives, and various other health care workers. While this forms a global problem, it should be noted that related research needs to be tailored to exploring solutions aimed at replacing the workforce actually using the technology. This could be in the form of diagnosis, monitoring, and delivery as well as finding ways to ensure these solutions can reach remote settings where access may not necessarily be available. As such, to ensure research becomes of value to society, it should aim to focus on linking health needs to appropriate innovations in medical devices.

1.4. RESEARCH AIMED AT IMPROVING MEDICAL DEVICES

Research aimed at improving the operation of medical devices is guided largely by the need for better solutions and greater technological capabilities. This, in addition, is fueled by promising ideas, scientific interest, and various economic factors. Much of this research is targeted to first world countries, which often fails to adequately address global public health needs. The main forces or drivers of medical device-related research arise from three main areas, namely academia, industry, and public research foundations.

1.4.1. Academia

While improving public health is one of the biggest drivers for scientists; their interest in discovering new knowledge, theories, and solutions that will aim at advancing scientific knowledge is also a major factor that must be considered. Most of the research associated with academia takes place in a university environment such as in medical or engineering faculties. Often research strategies carried out at academic institutions are typically driven by the research interests of professors as well as the likability of a proposed research project with surrounding academic opinions from various people. In addition, the ability to acquire funding or research grants will also influence the research interests that are primarily focused on. Traditionally, academic research receives most of its funding from governments, research foundations, and industrial and/or commercial institutions who in turn, exert a specific influence on the research protocols and objectives set out.

Ideas from researchers, scientists, clinicians, and engineers are normally submitted for peer review by reviewers who are experts in that field. The reviewers then base their assessments on the scientific credibility of the ideas, originality, and the extent to which they are innovative. The gap, however, may lie in transferring knowledge from academia to industry due to differences in schedules, expectations, outcomes, and overall working culture in conjunction with differences in communication and lack of ability to collaborate in certain instances.

1.4.2. Industry

Medical devices are often developed via a bottom-up approach whereby manufacturers explore various ideas for developing new products or ways in which they can improve existing ones. This is done in conjunction with health professionals and other device users and often takes into consideration various economic, political, and social factors. Research that is funded by industry is generally applied to specific industry targets and rarely goes beyond the market focus of the funder themselves. Large manufacturers tend to have their own funds for research, while smaller firms tend to depend on external sources of capital. Regardless, research and development spending can amount to extreme amounts, and as such, these firms must take measures to control or curb unnecessary or exorbitant costs.

Investment in research and development within the medical device industry happens for three main reasons, namely: (a) the likelihood of returns on R&D investment; (b) the status of R&D as a core element of a company's business model; and (c) the need to strengthen and expand the presence of a company in various markets. Major medical device manufacturer spends on average around 7.5% of their sales revenue on R&D (Table 1.1).

Another driver of research and development in industry is the progress being made in enabling technologies in those disciplines that are not directly related to medical devices. These include materials science, energy management, and so on

Table 1.1: Global Industry R&D Investment in Comparison to Revenue from Sales

Rank	Manufacturer	Headquarters	Revenue from Sales (US$ Millions)	R&D Investment (US$ Millions)
1	Boston scientific	United States	8,050	1,006
2	St. Jude Medical	United States	4,363	532
3	Siemens Healthcare	Germany	15,526	1,630
4	Philips Healthcare	Netherlands	9,227	892
5	Medtronic	United States	13,515	1,275
6	Beckman Coulter	United States	3,099	280
7	C.R. Bard	United States	2,452	199
8	Johnson & Johnson	United States	23,225	1,858
9	Drager	Germany	1,729	136
10	Olympus	Japan	3,920	289
11	Alcon	Switzerland	2,881	204
12	Baxter International	United States	12,400	868
13	Varian Medical	United States	2,070	136
14	Hospira	United States	3,620	214
15	Danaher Corporation	United States	3,227	190
16	Becton Dickinson	United States	7,156	396
17	Stryker	United States	6,718	368
18	Synthes	United States	3,206	170
19	Terumo	Japan	3,400	175
20	Zimmer	United States	4,121	194
21	Smith & Nephew	United Kingdom	3,801	152
22	B. Braun	Germany	5,263	181
23	Covidien	Ireland	9,910	340

24	Fresenius Medical Care	Germany	2,874	80
25	Dentsply	United States	2,194	52
Total			**157,947**	**11,817**

Adapted from: WHO, 2010-Medical Devices: Managing the Mismatch: An Outcome of the Priority Medical Devices Project.

1.4.3. Governments and Public Research Foundations

Governments and other public institutions tend to invest in research due to the proven benefits to public health. Much of this funding tends to promote innovation at universities and other publicly funded institutions. This funding can be geared towards covering research on medical devices that have scientific merit rather than on the magnitude of the public health burden that the devices might help to reduce. One example of this would be The German Research Foundation. The European Commission urges its high-income Member States to spend about 3% of their gross product on the development of cutting-edge technologies that would promote innovation and not focus specifically on a public health issue. While that does have some benefits, without explicit reference to public health needs, the political and economic motivation to invest in advanced technologies can shift priorities.

1.5. CHOOSING A MEDICAL DEVICE

The process of choosing a medical device is a complex one and often requires a transparent process that is based on information, reasoning, evidence, assessment of public health needs and a process of priority. The need to improve individual and public health must always be taken into consideration. If this is not done, it can lead to inappropriate use and an overall waste of financial resources. The above-mentioned factors tend to have negative consequences in both industrialized and developing countries. One such example is diagnostic imaging. The WHO estimates that high-tech diagnostic imaging is required only in 20–30% of medical cases worldwide. Of those cases, around 80–90% of diagnostic problems can generally be solved using basic x-ray and/or ultrasound examinations.

With the multitude of manufacturers available worldwide, each of whom produce various kinds of medical devices on a daily basis; it is understandable that choosing an appropriate device that is safe and which performs as intended has become difficult for ministries of health, regulatory authorities, managers, and health care facilities. In addition, the lack of professional technical advice from medical device specialists also contributes to these difficulties. There are various barriers to choosing medical devices, such as fascination with technology, aggressive marketing, high costs, and inadequate information received or known about the device.

1.5.1. Lack of Information

The act of purchasing a medical device is based on possible needs that the device will fill, and as such, the rationale for choosing a medical device will therefore call for information about the needs and extent to which a device will meet that need. A lack of adequate information may result in a failure to adopt a rational or logical assessment of needs which in turn may form one such barrier when it comes to choosing suitable devices that would aid a positive outcome in health care. Much of general healthcare spending is tailored to activities that do not actually focus on improving healthcare or better understanding the advantages among various choices, which has led to an overall gap in knowledge. This ultimately leads to a lack of knowledge or high-quality data, which in turn may result in incorrect choices being made about certain devices carrying a significant degree of risk. A common example is hospitals that choose to offer cutting-edge technology such as proton therapy which is known to be quite expensive. While this treatment is new and advanced, there is no evidence to suggest improved clinical outcomes when compared to conventional treatment. These scenarios often stem from a lack of objectivity and information availability. There are several sources of information that can assist health care practitioners. Regulatory bodies (such as the FDA in the United States and the Medicines and Healthcare products Regulatory Agency (MHRA) in the United Kingdom) provide information about medical devices however this information is generally limited to that needed for market approval and concerns around safety and efficacy of performance both, pre-, and post-market. It should be remembered that the main function of regulatory authorities is to ensure that a device complies with statutory requirements for safety, performance, and efficacy.

A second factor limiting information availability is that innovative medical devices are usually evaluated for efficacy in clinical trials. These trials however are too small to identify uncommon complications. In addition, randomized controlled trials tend to be impracticable or even unethical in the case of medical devices. One of the results of limited information availability is that estimates around efficacy or, rather, the cost-effectiveness of innovative technologies become quite vague. Where pre-market clinical trials are concerned, information availability may also become restricted. This occurs as clinicians participating in a pre-market trial tend to be well trained and alert to possible limitations. This is in direct contrast to what the requirement should be-clinicians participating in trials should be unbiased and unaware of the device's shortcomings.

1.5.2. Attractive Technology

Often fascination with science and technology can blind decision-makers and influence their ability to make objective decisions. In more cases than not, it is the existence of a technology that tends to be more appealing than the actual contribution that the technology has to overall patient care. In recent years many health care practitioners have become reluctant to purchase expensive medical equipment. In fact, national health systems, private insurers, and the like are now demanding that innovation be linked to economic value and focus on improving health outcomes.

1.5.3. Personal Preference

As time passes, many health care professionals tend to develop personal preferences for specific brands of medical devices. These preferences are generally for brands that they are familiar with or use often and which are often critical to the success of a clinical procedure. It should also be noted that the actual skill of a clinician themselves when using a medical device can also influence the outcome of a procedure which in turn could affect the overall preference towards a device. In addition, the special relationships that clinicians may develop with manufacturers of brands should also be taken into consideration as these may affect the objectivity and professional integrity of the professional concerned. Often these preferred items may include quite expensive devices that, in turn, may account for more than 50% of a hospital's expenditure. Such devices include implantable pacemakers, stents, and so on. Much of these personal preferences can also influence the choice of medical devices chosen in less developed countries. One such example is students who leave poorer countries to study medicine in developed countries like the United States or Europe. During this period, they may learn to work with expensive and complex medical devices while abroad, but when returning home, they may not have the same facilities available to them. Therefore, they may put motivations or pressure for the hospital to purchase expensive equipment without taking into consideration the actual public health benefit it could bring.

1.6. COSTING OF MEDICAL DEVICES

The rising costs associated with healthcare form a cause of concern to a growing number of countries. This is especially the case with countries that have aging populations and, as a result, increased demands for healthcare. The majority of industrialized countries have cost-containment measures in place for medicines, but very few have such measures in place for medical devices. In addition, low-income countries have their own problems. These include extremely high costs associated with expensive equipment, which in turn could strain the health system budget. Furthermore, the cost of acquiring a complex medical device is usually only one portion of the financial burden. Accessories to the device, years of warranty, delivery logistics, installation procedures and maintenance costs, spares, consumables, and reagents all need to be taken into consideration (Box 1.1). All these additions can, in actual fact, amount to more than 80% of the total cost of the device itself.

Box 1.1: The Hidden Costs of Medical Devices

- Service contracts;
- Spare parts;
- Test equipment;
- Depreciation;
- Downtime;
- Space;
- Licenses;
- Upgrades;
- Product management;
- Labor;
- Utilities;
- Training;
- Supplies;
- Installation;
- Accessories.

Inconsistency of pricing can lead to further complications during the procurement process, while improper cost management systems can pose overall difficulties for some countries. Often, the pricing between public health care facilities and private health care facilities differ drastically, which is an additional factor accounting for inconsistencies.

1.6.1. Lack of a Single Nomenclature

A lack of standardization tends to negatively affect the choice of medical devices chosen. Particularly lack of standardization in three main areas, namely regulations, standards, and nomenclature can severely affect procurement activities. There are several different nomenclatures being used throughout the world, namely the GMDN, the UMDNS, the Standard ISO 9999 and the Harmonized System of the World Customs Organization. In addition, other nomenclatures have been designed by individual countries such as Japan, Mexico, Norway, and so on; specifically, for their own specific purposes. Each nomenclature system identifies a given medical device by a name and/or code, which in turn defeats the main purpose of a single universal nomenclature system. Having a universal nomenclature system could lead to several benefits. Firstly, the presence of several different nomenclature systems around the world could make it extremely difficult for procurement managers, health officials, national regulatory authorities, hospital inventory managers, marketing managers and so on to exchange information about a medical device, for example, information surrounding adverse event reporting which often is exchanged across various countries.

1.6.2. Solutions to Overcoming Barriers Experienced When Choosing a Medical Device

While there are a variety of solutions to overcoming respective barriers, this section will focus on the main areas that could be applicable to or involved in any possible solution. This includes rationale decision making, public health needs, cost containments and improvement of existing marketing practices. Often regulatory practices tend to differ from one country to the next. Much of this diversity originates from differences in historical and cultural practices. One of the main objectives of the GHTF is to merge or harmonize differing regulatory practices. This is achieved via various guidelines and recommendations produced by the GHTF, which essentially, when focused on collectively; form a framework that any country can use to align their regulatory systems with that of other countries. Currently, there are few standards against which compliance with reliability, safety, efficiency, and quality of medical devices can be measured. International standards have been developed and put forward by several international organizations such as the International Organization for Standardization (ISO), The International Electrotechnical Commission (IEC), and the International Telecommunication Union; all of which are now widely used within the medical device community.

1.6.3. Common Questions That Could Be Asked When Choosing a Medical Device

Rational processes should be employed when making decisions about which medical devices to obtain. These processes should consider public health needs, cost-effectiveness, and overall health outcomes (Box 1.2).

Box 1.2: Common Questions That Could Be Asked When Choosing a Medical Device

- What are the clinical needs that the device is aimed at fulfilling?
- Does the need correspond to accepted healthcare practices?
- Will the device improve public health?
- Is the cost of the device justified by the need?
- Will paying the cost of the device divert funds needed for other, high er-priority needs?
- Is it a priority in this particular setting?
- Has evidence of adequate safety and effectiveness been obtained?
- Is the device appropriate to the level of health care (primary, second ary, tertiary) at which it will be used?
- Is the device appropriate to the likely availability of resources for up keep, maintenance, repair, and other ancillary requirements?
- Is training being provided?
- Is the physical infrastructure adequate?

When choosing a medical device, a health care facility or public health authority should make the decision based on answers to various, relevant questions as identified above. These questions are best answered by a team consisting of several specialists within the health care or scientific fields. These could include individuals such as biomedical engineers, medical staff, nursing staff, and even financial staff; all of whom have relevant knowledge and who are mandated to make choices that focus on the health needs of a target population.

The size and composition of the team will differ depending on the size and nature of the healthcare facility. Furthermore, clinical guidelines could be more useful. A guideline could take various forms, such as a flowchart or decision algorithm which ideally guides the user through a series of steps or choices that ultimately identifies a list of suitable devices needed for the management of a given disease or disability. These protocols should, ideally, urge readers to consider the required technical skills and supportive infrastructure needed and which is available for safe and effective use of the recommended device. These guidelines or protocols could be used at any level of the health system; from district hospitals to provincial hospitals to regional hospitals; as well as for any clinical conditions ranging from mild to severe.

1.6.4. Assessment of Costs

When it comes to assessing prices of medical devices, procurement teams may rely on consumer reports, supplier catalogs, and various other information sources. To further improve processes and limit costs, a procurement team could benchmark medical device prices or develop what is known as a 'formulary' of preferred suppliers. The aim of this list is to limit marketing pressure usually from large numbers of suppliers. In addition, pooled purchasing or bulk procurement arrangements may be used to limit costs. This is often the case in some countries or international agencies. Reference price lists of medical devices used in selected countries could be made available to the public in order to facilitate price benchmarking. A list of key medical devices used in the management of certain high-burden diseases could serve as a reference product list which in turn could be used for price benchmarking.

1.6.5. Improving Marketing Practices

There are a range of solutions that could be used to improve marketing practices. These include improving regulations, tackling corruption, and dealing with counterfeiting production. Despite the existence of regulations in many countries, the actual implantation tends to be highly variable in overall efficacy and comprehensiveness. The ideal scenario would be that no medical device should be used without having acquired market approval from a fully operational regulatory system. This, however, requires every country without a regulatory system to begin putting such frameworks in place. Furthermore more, protocols could be developed that allow devices regulated in one country with a sound established regulatory system to be automatically regulated or deemed to fulfill regulatory requirements in countries not yet possessing a fully operations regulatory system. Corruption is also a problem however attempting to combat corruption-related issues may serve in reducing associated costs. The availability of information or rather, transparency of information could assist in certain healthcare facilities reducing the amount paid for certain devices. Procurement guidelines, codes of conduct for health sector operators, and overall transparency and monitoring of procedures could assist in reducing corruption and lowering overall costs.

Chapter 2

Medical Device Use and Regulation

CONTENTS

2.1. USING MEDICAL DEVICES

Once a proper choice of medical device is made, the next area for concern is whether or not that device will be properly used and if not, why. For a device to be used, it must be "usable." The main barrier to the actual usability of a medical device is the inappropriateness of its design in relation to its intended use. As such, the responsibility should fall on the manufacturer to determine the context in which a device will be used as well as to design their devices to be contextually appropriate. There are, however, various issues that may serve as barriers to medical device use. These factors include geographical, social, cultural, economic, demographic, and medical factors, all of which influence where and how a medical device will be used.

2.1.1. Inappropriate Design

Contextual factors should be considered when designing medical devices to ensure they are adequately constructed. Understanding the context of use is important as this often forms a complex factor that influences the use of the device in everyday working environments. There are four main layers of contextual factors, namely: (a) health care facility characteristics; (b) supplies and expertise; (c) organizational structure; and (d) expectations (Figure 2.1). Each of the top layers is dependent on the layer below it.

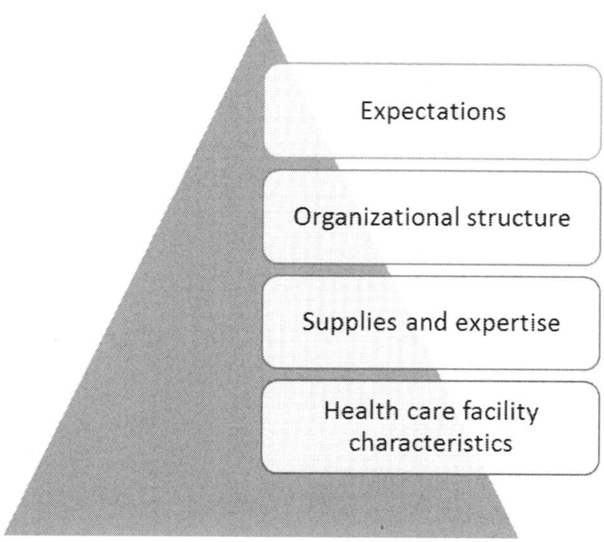

Figure 2.1: A context pyramid outlining factors affecting the design of medical devices.

For example, the overall value is minimized if the focus is first on patients' expectations before addressing adequate training of the healthcare personnel that will be operating the device. It should also be noted that contextual factors can be interpreted from a variety of perspectives such as that of the user, patient, manufacturer, and so on. the user's perspective is important for the design of the device as well as the health care facility in which it will be operated. For example, devices to be used in a primary health care facility will differ from devices to be used at home in respect to how they will be operated and safety/operational instructions (Table 2.1). The use of such devices in a home-based setting is becoming more and more apparent, and as such, this factor should be taken into consideration when manufacturers are designing and developing their products. Patients in a home-based setting are far less skilled than professional individuals within a health care setting. In this regard, such patients are less able to overcome various limitations of the device, and it should be taken into consideration when developing the risk assessment of the product.

Table 2.1: Examples of the Purpose, Place of Use and Users of Common Medical Devices

	Medical Devices			
Context of Use	**Preventative**	**Diagnostic**	**Therapeutic**	**Assistive**
Main users	Healthcare professionals or healthy individuals	Healthcare professionals or patients	Healthcare professionals or patients	Individuals or healthcare professionals
Examples of medical devices used in healthcare facilities	Surgical gloves, sterilization equipment, disinfectants	Laboratory diagnostic tests, X-ray equipment, MRI, electrocardiogram, stethoscopes, blood gas analyzers, endoscopes, tongue depressors, reflex hammers	Orthopedic implants, surgical equipment, pacemakers, stents, infusion pumps, ventilators, sterile dressings, laparoscopes	Traction devices, patient hoists, hospital beds, operating tables, prostheses
Examples of medical devices used in homes	Condoms, gloves, pessary	Pregnancy tests, blood glucose tests, blood pressure meters, telemedicine, cardiac monitoring	Infusion pumps, dialysis equipment, oxygen supply systems, syringes	Crutches, wheelchairs, spectacles, eye lenses, hoists

Resource settings also play a factor in the design of the device. Low resource settings often present a challenge for manufacturers and governments when implementing and attempting to meet certain medical device regulations as well as applying these regulations to improve and maintain country health. Most medical equipment in low-resource countries tends to be imported from developed or high-resource countries and, upon arrival, meet various challenges when attempting to be put into operation. One of the biggest reasons for this is the disparity between a low-resource and high-resource countries and the context in which they operate. One of the most common examples is the presence of stable sources of electricity and clean running water in industrialized countries. Often these are taken for granted in industrialized countries where they are needed for the efficient and correct operation of certain medical devices. However, this is not always the case in low resource countries. As a result, manufacturers have had little reason to take a lack of electricity and water into consideration when designing and producing devices that will have to function in parts of the world where such resources are scarce or in some instances, non-existent. It should also be noted that the design of some medical devices can also be problematic in high resource settings. For example, through faculty operator-interface designs, inappropriate packaging, or incorrect packaging.

2.1.2. Limited Management

The procurement of many medical devices happens without having a proper medical device management plan in place. These types of plans are essential to showing how to maintain medical device functionality, safety, accuracy, and durability. The management of medical devices in both, high-, and low-resource settings can be difficult. High-resource settings may see the presence of multiple similar devices with variations in model types, for example. This in turn would require a need for overall standardization of processes when it comes to operation, use, repair, servicing, and so on. The lack of such standardization could form a barrier to efficient operation. In low resource settings, however, many of these devices are out-of-order or operating with poor functionality. This is often due to limited management in the form of lack of training of operations and staff amongst the lack of other essential services. This could also be due to lack of proper infrastructure such as electricity, internet connections, and landlines; all of which are essential for medical devices to be operated to their full potential. In addition, these countries often lack funds and expertise to create and run efficient medical device management systems.

Management of medical devices should be all-round regardless of the risk class and complexity of the device. It is a common misconception that only high-risk devices require suitable management plans, but even basic commonplace devices such as disposal plastic syringes should have suitable management plans detailing how they should be disposed of and what the various safety instructions are during use.

2.1.3. Lack of Training

The efficient operation of any medical device, regardless of the risk classification or complexity requires knowledge and skills; all of which are acquired through education, training, and experience. Some devices may require well-coordinated, well-trained teams of specialists, professionals, and laboratory backup staff for their operation, while others may require short training videos and active online portals. Regardless, ensuring there is adequate transfer of knowledge and skills is essential. The lack of the above-mentioned may however form a considerable barrier to the safe and efficient use of medical devices. The lack of adequate training often leads to a high frequency (HF) of reported errors in the use of medical devices. For example, in 2000, the FDA received over 90,000 reports on device-related errors.

There are, however several constraints experienced during the arrangement of adequate training of medical devices. The first is the timeframe involved. Often the acquiring skills needed to use more complex medical devices involves a particularly long time to understand and transfer knowledge. In addition, the turnover time for trainings is often quite frequent due to the short commercial lifecycle of medical devices. Often this is around on average 18 months before new, more technologically advanced models are released to replace earlier models. Working conditions may also hamper proper training such as lack of facilities and materials as well as the organization of training sessions for part-time employees who may work for a few days a week or only night shifts thereby causing them to miss trainings.

Furthermore, one-time training sessions are often insufficient to meet the level of competency needed to operate more complex devices. In some instances, the presence of technical information, in the form of instruction manuals or direct communication from distributors, could serve as a substitute for actual trainings, which could be disastrous. In addition, many of these manuals may not be in the language of the receiving country, making it difficult to conduct trainings or even begin understanding how to operate the device. Even the few manuals that are translated into a local language may not be done adequately and might contain unreliable information due to inaccuracies of translation. In some instances, training curriculums have been devised, but these can very easily become out-of-date. In some poorer countries, suitably qualified candidates tend to leave to work in more industrialized countries which offer more favorable career prospects and more opportunities to acquire skills in using high-tech devices, thereby further contributing to the problem.

2.1.4. Maintenance Problems

One of the biggest reasons for medical devices not being in operation is lack of, or inadequate maintenance. This is often a problem in developing countries where there is a lack of laid out budget, suitable industry technicians and ready access to spare parts.

The need for suitable industry technician and properly trained staff is essential for proper installation and to ensure a medical device is kept in good working condition. In addition, one device may have different parts that require different types of maintenance procedures and maintenance schedules.

A lack of appropriate consumables is also a problem. As these can only be used once or for a limited time and then needs to be discarded and replaced. Common examples of consumables include test strips, electrodes, operating gowns, sterilizing liquids, dressings, reagents for diagnostic equipment, and so on.

The lack of standardization in maintenance procedures and schedules can often lead to a poor search for suitable consumables that are compatible with specific medical device models. The need to acquire consumables can however present a problem to low-resource countries due to ongoing costs and storage requirements.

2.2. HOW ARE MEDICAL DEVICES REGULATED IN EUROPE?

Medical devices cannot be placed on the European market without a manufacturer proving conformance to the strict safety requirements of the European Union which, once done, allows for the affixation of the CE conformity mark on the device itself. These safety requirements are laid out in relevant regulations, previously known as the MDD and now the MDR. The MDD formed the core legal framework in Europe since the 1990s. It consists of three directives that regulate the safety and marketing of medical devices. These three directives are:

- Active implantable medical device directive (AIMDD 90/385/EE);
- Medical device directive (MDD 93/42/EEC); and
- *In vitro* diagnostic medical device directive (IVDMDD 98/79/EC).

These have been supplemented since by several necessary updates and relevant guidance documents and directives to guide manufacturers when implementing the regulation. Often these guidance documents and directives are released due to new and emerging technologies which challenge current frameworks or highlight gaps or areas of confusion.

The MDD defines a medical device as:

"Any instrument, apparatus, appliance, material or other article, whether used alone or in combination, including software necessary for its proper application intended by the manufacturer to be used for human beings for the purpose of:

- Diagnosis, prevention, monitoring, treatment, or alleviation of disease;
- Diagnosis, monitoring, treatment, alleviation of or compensation for an injury or disability;
- Investigation, replacement, or modification of the anatomy or of a physiological process;
- Control of conception and which does not achieve its principal intended action in or on the human body by pharmacological, immunological, or metabolic means, but which may be assisted in its function by such means."

The CE mark is a conformity mark which all European medical devices must have before they can be placed on the European market. By placing this mark on a device, it is seen as a declaration by the manufacturer that the product meets all the appropriate provisions of the relevant legislation, including those related to safety and, where required, has been assessed in accordance with these.

Article 17 (CE Marking) of the MDD 93/42/EEC, States:

1. 'Devices, other than devices which are custom-made or intended for clinical investigations, considered to meet the essential requirements (ER) referred to in Article 3, must bear the CE marking of conformity when they are placed on the market.
2. The CE marking of conformity, as shown in Annex XII, must appear in a visible, legible, and indelible form on the device or its sterile pack, where practicable and appropriate, and on the instructions for use (IFU). Where applicable, the CE marking must also appear on the sales packaging.

   It shall be accompanied by the identification number of the notified body responsible for implementation of the procedures set out in Annexes II, IV, V, and VI.
3. It is prohibited to affix marks or inscriptions which are likely to mislead third parties with regard to the meaning or the graphics of the CE marking. Any other mark may be affixed to the device, to the packaging or to the instruction leaflet accompanying the device, provided that the visibility and legibility of the CE marking is not thereby reduced.

The CE conformity mark consists of the initials 'CE' and is usually represented as in the figure below:

CE

There are specific requirements are the visual display of the CE mark as well, including that the various components of the CE mark must have the same vertical dimension, which may not be less than 5 mm (although this minimum dimension may be waived for small-scale devices). It is also important to note that the CE marks or rather the letter "CE" does not stand for any specific words but rather it is a symbol indicating a declaration by the manufacturer that the product meets all the appropriate provisions of the relevant legislation.

It should not be placed on a device if the manufacturer cannot prove conformity to the regulation and that the device has been assessed in accordance with the appropriate procedures. The CE mark also means that the product can be freely marketed anywhere in the European Economic Area (EEA) without further control.

2.3. OVERVIEW OF THE CE MARKING PROCESS

There are a range of regulatory bodies involved in the CE marking process. These include competent authorities (CA), NBs, and the authorized representatives (AR). It is important for a manufacturer to understand the difference between each one. These are explained in subsections.

2.3.1. Competent Authorities (CA)

A competent authority is defined as a regulatory body authorized by a European Member State government to monitor compliance of devices with the national legislation or regulations and carry out duties on behalf of the government in compliance with European Law. CA's existed in each European Member State and is nominated by each government to monitor and ensure compliance with the MDD (and upcoming MDR). The main role of the CA is to watch over the market in the member state for which it is designated. In particular, it must advise about the market launch of medical devices, advice about exporters and distributor's performance, advice about any clinical studies being done with medical devices in that particular territory, and advice about incidents that occur with medical devices. The CA designates a NB to ensure that conformity assessment procedures are completed as per relevant criteria.

A list of National CA within the EU is provided in Box 2.1.

Box 2.1: National Competent Authorities within the European Union

Austria:
Federal Ministry of Health-Department, Pharmaceuticals, and Medical Devices, *Radetzkystrasse – 2,1030, Wien/Vienna, Austria,* ***E-mail:*** *meddev@bmgf.gv.at*
Federal Office for Safety in Health Care, *Traisengasse – 5,1200, Vienna, Austria,* ***Tel.:*** *+43-50-555-36402,* ***E-mail:*** *medizinproduckte@basg.gv.at,* ***Website:*** *www.basg.gv.at*
Belgium:
Federal Agency for Nuclear Control, Health Protection Unit, *Rue Ravenstein – 36, 1000, Brussels,* ***Tel.:*** *+32-2-289-21-11,* ***Fax:*** *+32-2-289-21-12,* ***E-mail:*** *point-contact@fanc.fgov.be,* ***Website:*** *www.afcn.fgov.be*
Federal Agency for Medicines and Health Products, Health Products Division, *Place Victor Horta 40/40, 1060, Brussels,* ***Tel.:*** *+32-2-524-80-00, Fax: +32-2-524-81-20,* ***E-mail:*** *meddev@afmps.be,* ***Website:*** *www.fagg-afmps.be*
Bulgaria:
Bulgarian Drug Agency, Department Control of Medicinal Products and Medical Devices *8, Damyan Gruev Str., 1303 Sofia, Bulgaria,* ***Tel.:*** *+359-2-890-34-83,* ***Fax:*** *+359-2-890-34-34,* ***E-mail:*** *bda@bda.bg,* ***Website:*** *www.bda.bg*
Croatia:

Ministry of Health, Health Protection Directorate, *Ksaver 200a, 10000 Zagreb, Croatia,* ***Tel.:*** *+385-1-4607555,* ***Fax:*** *+385-1-4677076,* ***Website:*** *www.zdravlje.hr*
Agency for Medicinal Products and Medical Devices, *Ksaverska Cesta – 4,10000 Zagreb,* ***Tel.:*** *+385-1-4884-300,* ***Fax:*** *+381-1-4884-110,* ***E-mail:*** *halmed@halmed.hr,* ***Website:*** *www.halmed.hr*

Cyprus:

Ministry of Health, Cyprus Medical Devices Competent Authority *1 Prodromou and 17 Chilonos Street, 1448 Nicosia, Cyprus,* ***Tel.:*** *+357-22605572/735,* ***Fax:*** *+357-22468427,* ***E-mail:*** *cymda@mphs.moh.gov.cy,* ***Website:*** *www.moh.gov.cy/mphs*

Czech Republic:

State Institute for Drug Control, *Šrobárova 48, 100-41 Praha 10,* ***Tel.:*** *+420-272-185-111,* ***E-mail:*** *posta@sukl.cz*

Denmark:

Danish Health and Medicines Authority, *Axel Heides Gade 1 DK – 2300 Copenhagen S,* ***Tel.:*** *+45-72-22-74-00,* ***Fax:*** *+45-44-88-95-99,* ***E-mails:*** *med-udstyr@dkma.dk; dkma@dkma.dk,* ***Website:*** *www.medicinskudstyr.dk*

Estonia:

Health Board, Medical Devices Department, *1a Põllust., EE-Tartu – 50303,* ***E-mail:*** *MSO@terviseamet.ee,* ***Website:*** *www.terviseamet.ee*

Finland:

National Supervisory Authority for Welfare and Health (Valvira), *Mannerheimintie 103 b, P.O. Box 210, FI – 00281 Helsinki, Finland,* ***Tel.:*** *+358-295-209-111,* ***Website:*** *www.valvira.fi*

France:

Ministry of the Economy and Finances, Direction Générale de la Concurrence, de la Consommation et de la Répression des Frauds-Unit 5B, *Télédoc 241-59, Boulevard Vincent-Auriol F – 75703 Paris Cedex 13,* ***Tel.:*** *+33-1-44-97-28-01,* ***E-mail:*** *Bureau-5B@dgccrf.finances.gouv.fr,* ***Website:*** *www.economie.gouv.fr/dgccrf*
Agence Nationale de Sécurité du Médicament et des Produits de Santé (ANSM), *143/147, Boulevard Anatole France – 93285 Saint-Denis Cedex,* ***Tel.:*** *+33-1-55-87-36-86/+33-1-55-87-37-28,* ***E-mails:*** *dmdpt@ansm.sante.fr; dmtcos@ansm.sante.fr,* ***Website:*** *www.ansm.sante.fr*
Direction Générale des Douanes et Droits Indirects, Tel.: *0811-20-44-44/+33-1-72-40-78-50,* ***E-mail:*** *ids@douane.finances.gouv.fr,* ***Website:*** *www.douane.gouv.fr*

Germany:

Federal Institute for Drugs and Medical Devices, *Kurt Georg Kiesinger Allee 3, D – 53175 Bonn, Fax: +49-228-207-5300,* ***E-mail:*** *medizinprodukte@bfarm.de*
Zentralstelle der Länder für Gesundheitsschutz bei Arzneimitteln und Medizinprodukten (ZLG), E-mail: *zlg@zlg.nrw.de*
Federal Ministry of Health, *Rochusstrasse 1, D – 53123 Bonn,* ***Fax:*** *+49-228-941-4946,* ***E-mail:*** *medizinprodukte@bmg.bund.de*

Greece:

National Organization for Medicines, *284 Mesogion Ave, GR – 15562 Holargos, Athens,* ***E-mail:*** *relation@eof.gr,* ***Website:*** *www.eof.gr*

Hungary:

Health Registration and Training Center (HRTC), *H-1085 Budapest, Horánszky street 15,* ***Tel.:*** *+36-1-302-5060, Fax: +36-1-269-1255,* ***E-mail:*** *amd@enkk.hu,* ***Website:*** *www.enkk.hu*

Iceland:

Medicines Agency, *Vínlandsleið 14, 113 Reykjavík,* ***Tel.:*** *520-2100,* ***Fax:*** *561-2170,* ***E-mail:*** *ima@ima.is,* ***Website:*** *www.imas.is*

Ireland:

Health Products Regulatory Authority (HPRA), *Kevin O'Malley House, Earlsfort Centre, Earlsfort Terrace, Dublin 2,* ***Tel.:*** *+353-1-676-4971,* ***E-mail:*** *info@hpra.ie,* ***Website:*** *www.hpra.ie*

Italy:

Ministry of Health, Department of Planning and Organization of the National Health Service-Directorate, General of Medical Devices, Pharmaceutical Services, and Safety in Healthcare, *via Giorgio Ribotta 5, IT – 00144 Roma,* ***E-mail:*** *segr.dgfdm@sanita.it,* ***Website:*** *www.sanita.it*

Latvia:

Health Inspectorate, *Klijanu Street 7, Riga, LV—1012, Latvia,* ***Tel.:*** *371-67819671,* ***Mob:*** *371-26315189,* ***Fax:*** *371-67819672,* ***E-mail:*** *vi@vi.gov.lv,* ***Website:*** *www.vi.gov.lv*

Liechtenstein, Amt Für Gesundheit, Äulestrasse 51, Postfach 684, 9490 Vaduz, Liechtenstein, E-mail: info.ag@llv.li, Website: www.ag.llv.li

Lithuania:

State Health Care Accreditation Agency, Jeruzales St. 21 LT – 08420 Vilnius Lithuania, ***Tel.:*** *(85) 261-5177,* ***Fax:*** *(85) 212-7310,* ***E-mail:*** *vaspvt@vaspvt.gov.lt,* ***Website:*** *www.vaspvt.gov.lt*

Luxembourg:

Ministère de la Santé, *Villa Louvigny-Allée Marconi L – 2120 Luxembourg,* ***Tel.:*** *+352-24785500,* ***E-mail:*** *info@ms.public.lu*

Malta:

Malta Competition and Consumer Affairs Authority, Technical Regulations Division-Market Surveillance Directorate, *Mizzi House, National Road, Blata l-Bajda, Hamrun, Malta HMR – 9010,* ***Tel.:*** *+356-23952000,* ***Fax:*** *+356-21242406,* ***E-mail:*** *info@mccaa.org.mt,* ***Website:*** *www.mccaa.org.mt*

Netherlands:

The Health Care Inspectorate, *Stadsplateau 1-3521 AZ Utrecht, Postbus 2680-3500 GR Utrecht,* ***Tel.:*** *+31-88-120-50-000,* ***E-mail:*** *meldpunt@igz.nl,* ***Website:*** *www.igz.nl*

Norway:

The Norwegian Medicines Agency, *PO Box 240 Skøyen, 0213 Oslo, Norway,* ***Tel.:*** *+47-22-89-77-00,* ***E-mail:*** *meddev-no@noma.no,* ***Website:*** *www.legemiddelverket.no*

Poland:
Office for Registration of Medicinal Products, Medical Devices and Biocidal Products, Department of Medical, Devices Surveillance, Vigilance, and Clinical Trials, *Al. Jerozolimskie 181C, 02-222 Warsaw, Poland,* ***Tel.:*** *+48-22-492-15-54,* ***Fax:*** *+48-22-492-11-09,* ***E-mail:*** *incydenty@urpl.gov.pl,* ***Website:*** *http://www.urpl.gov.pl/*
Portugal:
INFARMED-National Authority of Medicines and Health Products, IP, *Parque de Saúde de Lisboa, Avenida do Brasil 53, 1749-004 Lisboa,* ***Tel.:*** *+351-217-987-100,* ***Fax:*** *+351-217-987-316,* ***E-mail:*** *infarmed@infarmed.pt,* ***Website:*** *www.infarmed.pt*
Romania:
National Agency for Medicines and Medical Devices, *58, Sos. Nicolae Titulescu, Sector 1, Bucharest,* ***Tel.:*** *+4-021-2228652; +4-031-432-9120,* ***Fax:*** *+4-021-2228683; +4-031-432-9121,* ***E-mail:*** *mdevice@anm.ro,* ***Website:*** *www.anm.ro*
Slovak Republic:
National Institute for Drug Control, *Kvetná 11-825-08 Bratislava 26 Slovak Republic,* ***Tel.:*** *+421-2-5070-11-11,* ***Fax:*** *+421-2-55-56-41-27,* ***E-mail:*** *sukl@sukl.sk,* ***Website:*** *www.sukl.sk*
Slovenia:
Agency for Medicinal Products and Medicinal Services, *Ptujska Ulica 21, 1000 Ljubljana, Slovenia,* ***Tel.:*** *+386 (0)8-2000-500,* ***Fax:*** *+386 (0)8-2000-630,* ***E-mails:*** *info@jazmp.si; meddev@jazmp.si,* ***Website:*** *www.jazmp.si*
Spain:
Agencia Española de Medicamentos y Productos Sanitarios, *C/Campezo 1, Edificio 8, ES – 28022 Madrid,* ***E-mail:*** *sgps@aemps.es,* ***Website:*** *http://www.aemps.gob.es/*
Sweden:
The Medical Products Agency, *P.O. Box 26, SE – 751-03 Uppsala,* ***Tel.:*** *+46-18-17-46-00,* ***Fax:*** *+46-18-50-31-15,* ***E-mail:*** *meddev.central@mpa.se,* ***Website:*** *www.lakemedelsverket.se*
The Health and Social Care Inspectorate, *P.O. Box 45184, SE – 104-30 Stockholm,* ***Tel.:*** *+46-10-788-50-00,* ***E-mail:*** *registrator@ivo.se,* ***Website:*** *www.ivo.se*
Switzerland:
Swissmedic, Swiss Agency for Therapeutic Products, *Hallerstrasse 7, CH – 3012 Berne,* ***E-mail:*** *medical.devices@swissmedic.ch,* ***Website:*** *https://www.swissmedic.ch*
Turkey:
Ministry of Health-Turkish Medicines and Medical Devices Agency, Vice Presidency of Inspectorate-Medical Devices Inspection Department, *Söğütözü Mahallesi 2176. Sokak No: 5 PK 06520 Çankaya/ANKARA,* ***Tel.:*** *+90-312-218-3534,* ***Fax:*** *+90-312-218-3710,* ***E-mail:*** *tibbicihazpgd@titck.gov.tr,* ***Website:*** *http://www.titck.gov.tr*

United Kingdom:
Medicines and Healthcare Products Regulatory Agency, *151 Buckingham Palace Road, London SW1W 9SZ,* ***Tel.:*** *+44-203-080-6000,* ***Fax:*** *+44-203-118-9803,* ***E-mail:*** *info@mhra.gsi.gov.uk*

2.3.2. Notified Bodies (NB)

A Notified Body is defined as an independent certification organization that is "notified" by the competent authority of a European Member State. It plays an imperative role when it comes to supporting a manufacturer's ability to place safe and compliant medical devices on the European market. The notified body is important for determining if a product or system meets the applicable requirements of the specific legislation/for CE marking and assists with carrying out conformity assessment procedures to demonstrate that the product complies with the requirements of the legislation. There are specific NBs dedicated to performing conformity assessments under Directive 93/42/EEC. However, it should also be noted that even fewer NBs are designated to perform conformity assessment under the Regulation (EU) 2017/745. Examples of these NBs are: BSI Assurance UK Ltd, BSI Group, the Netherlands B.V., DEKRA Certification B.V, TUV SUD Rheinland and Intertek Medical Notified Body AB to name a few.

2.3.3. Authorized Representatives (AR)

The European AR (also known as the EU AR, EC REP, CE REP, EU REP, or EAR) serves as a liaison between a manufacturer and the national competent authority and NBs. The AR should be identified on the product labeling, outer packaging, and the instructions for use (IFU) and will assist with device registrations. The AR will hold a copy of the manufacturer's technical file and/or CE Declaration of Conformity and will make it available to the CA upon request. They will also assist with Incident and Field Safety Corrective Action reporting. In many cases, the AR will also provide manufacturers with regulatory news and relevant updates.

The CA, NBs, and AR are all involved in the CE marking process which is interlinked between all three. The CA designates a NB to ensure that conformity assessment procedures are completed as per the relevant criteria. The authorized representative is designated by manufacturers and is legally responsible for compliance with the regulations. The AR acts as the first point of contact for the EU authorities. It is up to the manufacturer to ensure that their product complies with the ER of the relevant EU legislation.

A general overview of the CE marking process is given in Figure 2.2.

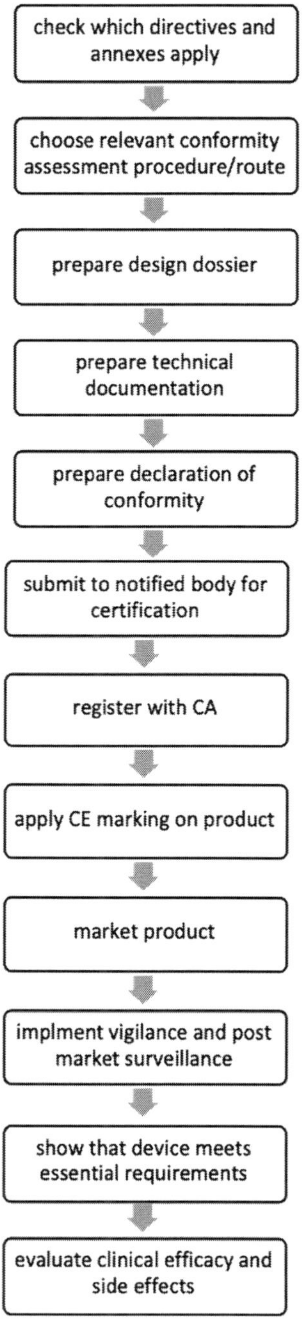

Figure 2.2: General overview of the CE marking process.

Chapter

3

Overview of the CE Mark Process

CONTENTS

3.1. SIMPLIFIED PROCESS FOR CE MARKING OF MEDICAL DEVICES IN EUROPE

3.1.1. Step 1: Determine If the Product Concerned Is a Medical Device

A medical device is defined as a product or equipment that is intended for use within the health care industry to either diagnose, monitor, or treat diseases or to act as a supportive aid for people with any form of illness or disability. There are over 500 000 different types of medical devices, and these include a vast array of devices such as x-ray machines, infusion pumps and cardiac pacemakers used by doctors, nurses, and surgeons as well as common devices that may be found every day in homes such as wound dressings, thermometers, contact lenses and wound irrigation solutions. According to ISO 13485, medical devices range from simple low risk products such as urine bags and latex gloves to more complex, technologically advanced programmable devices such as pacemakers. Generally, these devices are grouped into four classes namely Class I, Class IIa, Class IIb and Class III; with Class I being the lowest risk and Class III being the highest risk (Figure 3.1) or Class A, B, C, D with Class A being the lowest risk and Class D being the highest risk.

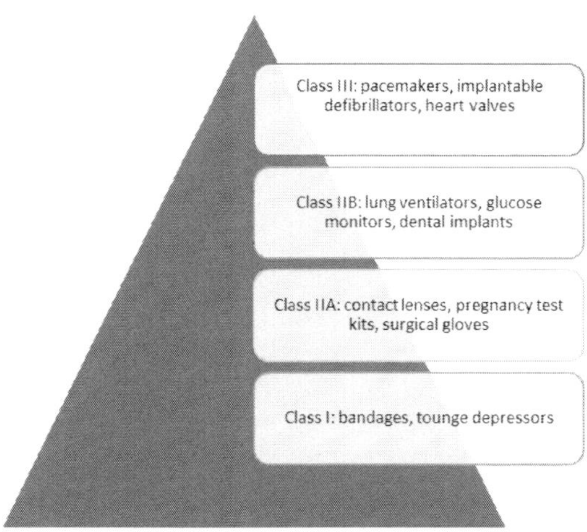

Figure 3.1: Overview of the different medical device classes.

The World Health Organization (WHO) defines a medical device as any instrument, apparatus, implement, machine, appliance, implant, reagent for *in vitro* use, software, material, or other similarly related article intended by the manufacturer to be used, alone or in combination and for human beings in one or more specific medical purpose(s) (Box 3.1). These devices should not achieve its main intended function by pharmacological, immunological, or metabolic means on or in the human body. It may however be assisted in its intended function by the above-mentioned methods.

Box 3.1: General Uses for Medical Devices

- Diagnosis, prevention, monitoring, treatment, or alleviation of disease.
- Diagnosis, monitoring, treatment, alleviation of, or compensation for an injury.
- Investigation, replacement, modification, or support of the anatomy and/or of a physiological process.
- Supporting or sustaining life.
- Control of conception.
- Disinfection of medical devices.
- Providing information by means of *in vitro* examination of specimens derived from the human body.
- Does not achieve its primary intended action by pharmacological, immunological, or metabolic means, in or on the human body, but maybe assisted in its intended function by such means.

Source: Derived from: WHO (https://www.who.int/medical_devices/full_deffinition/en/).

The mode of action of a medical device on the human body is what differentiates it from medicines, whose mode of action may often be metabolic, immunological, or pharmacological.

In fact, definitions must make clear distinctions on the difference between a medical device and medicine/drug whose primary intended use is obtained through chemical action or by being metabolized in the body. While many regulatory bodies tend to establish their own definitions around this topic, the Global Harmonization Task Force (GHTF) has proposed a harmonized definition for medical devices and *in-vitro* medical devices to ensure uniformity of understanding on a global scale (GHTF document SG1/N029R11) (Box 3.2).

Box 3.2: Definition of a Medical Device as Defined by GHTF (GHTF/SG1/N071:2012)

"Medical device" means any instrument, apparatus, implement, machine, appliance, implant, *in vitro* reagent or calibrator, software, material or other similar or related article, intended by the manufacturer to be used, alone or in combination, for human beings for one or more of the specific purposes of:

- diagnosis, prevention, monitoring, treatment, or alleviation of disease;
- diagnosis, monitoring, treatment, alleviation of or compensation for an injury;
- investigation, replacement, modification, or support of the anatomy or of a physiological process;
- supporting or sustaining life;
- control of conception;
- disinfection of medical devices;
- providing information for medical purposes by means of *in vitro* exami nation of specimens derived from the human body; and
- which does not achieve its primary intended action in or on the human body by pharmacological, immunological, or metabolic means, but which may be assisted in its function by such means.

The definition laid out by the GHTF covers a multitude of different medical devices, some of which are complex and reflects the latest advances in technological developments such as imaging equipment and implants; as well as those simple devices such as tongue depressors, thermometers, scales, and latex gloves. The GHTF document further defines those produces which may be considered to be medical devices in some jurisdictions while not classified in others. These include 1) disinfection substances, 2) aids for persons with disabilities, 3) devices incorporating animal and/or human tissues, and 4) devices for *in-vitro* fertilization or assisted reproduction technologies (Box 3.3).

Box 3.3: Definition of an *In-Vitro* Medical Device as Defined by GHTF (GHTF/SG1/N071:2012)

'*In vitro* diagnostic (IVD) medical device' means a medical device, whether used alone or in combination, intended by the manufacturer for the *in-vitro* examination of specimens derived from the human body solely or principally to provide information for diagnostic, monitoring or compatibility purposes.

3.1.2. Step 2: Classification of the Medical Device

Once it has been determined that the product is a medical device, the next step is to determine what its classification is. A medical device may be classified as:

- Class I-non-sterile, non-measuring;
- Class I sterile, measuring;
- Class IIa;
- Class IIb;
- Class III.

Class III medical devices cover the highest risk products. It should also be remembered that the higher the classification, the greater the level of assessment required. Classification rules are found in Annex IX of the MDD 93/42/EEC Directive and Annex VIII of the MDR 2017/745. The classification annexes include definitions of the terminology being used in the classification rules.

Classification of a medical device will depend on a series of factors. This includes:

- How long the device is intended to be in continuous use;
- Whether or not the device is invasive or surgically invasive;
- Whether the device is implantable or active;
- Whether or not the device contains a substance.

The class of the medical device is determined by the manufacturer's assigned intended purpose for use. It is important to remember that the classification should not be assigned based on the classification of other similar products manufactured by the same manufacturer or different manufacturers. For example, two sutures manufacturers by different manufacturers may well have the same composition but different intended purposes for use and hence different classifications. For this reason, a manufacturer may successfully change the classification of a product by clearly defining on the label the intended purpose in such a way that the device falls into a higher or lower class.

3.1.3. Step 3: Quality Management System (QMS)

The next step for a manufacturer is to implement a quality management system (QMS) in accordance with the medical device directive (MDD)/ medical device regulation. Most companies apply the ISO 13485 standard, which defines the QMS requirements for regulatory purposes. This standard is used for QMSs where an organization needs to demonstrate its ability to provide medical devices and related services that consistently meet customer and applicable regulatory requirements. It is important to note that compliance to ISO 13485 is not mandatory in all countries due to the presence of territory specific regulations. Nevertheless, certification in accordance with ISO 13485 often presumes compliance with certain annexes of the European MDD and now medical device regulations. In addition, depending on the class of the medical device in question, the requirements of the manufacturers QMS can be determined. For example:

- **For Class 1 Non-Sterile, Non-Measuring Devices (such as Examination Gloves):** Manufacturers do not need to comply fully to a QMS, i.e., they do not need not be audited to a full QMS. This is a self-declared conformity.
- **For Class 1 Sterile, Class IIa, Class IIb and Class III Medical Devices:** Manufacturers must operate under a QMS. Implementation of a QMS may take between 3 and 6 months and must be audited by a Notified Body which has been selected by the manufacturer.

3.1.4. Step 4: Technical File

While the implementation of a QMS is in progress, the manufacturer may choose to prepare the technical file of the devices which they are manufacturing. The technical file provides detailed information demonstrating compliance with the MDD 93/42/EEC and now, Medical Device Regulation 2017/745. The technical file is required for all medical devices, including Class I devices however the name and detail required for a Class III device will differ. For Class III devices, it is usually known as a design dossier. The technical file and design dossiers are audited by independent companies, which are better known as notified bodies (NB). These NB have been authorized by the EU Member State Ministry of Health. The manufacturer will select its own Notified Body for review of its technical files for the audit of the implemented QMS.

3.1.5. Step 5: Appoint an Authorized Representative

Manufacturers selling into the EU are required to appoint an Authorized Representative who will represent them in Europe, if their business is not located in Europe itself. An authorized Representative is also known as the

EC Rep. the EC Rep must be qualified to handle regulatory matters and will serve as the regulatory liaison between the Manufacturer and the European Ministry of Health of the respective European country, better known as the Competent Authority. Regardless of the classification of the medical device, manufacturers are required to appoint an EC Rep located in Europe. The ER Rep may also carry out other responsibilities as defined in the medical device regulations and as delegated by the Manufacturer. Tasks may include registration of medical devices, incident reporting, assistance with recalls and communication with the Competent Authority. The EC Rep will have full access to the Technical File or Design Dossier depending on the class of the device if a review is requested by the Competent Authority. The name and address of the EC Rep must appear on the labeling. It is also possible for manufacturers to consider appointing distributors to undertake the role of EC Reps. In this case, manufacturers may carefully choose the EC Rep base on capability and mandatory responsibilities. Distributors must be aware that the Technical Files, which is a proprietary document, must be made available when requested; however this can present problems with multiple Distributors in many different countries.

3.1.6. Step 6: Audit of the QMS and Technical File/Design Dossier

The QMS implemented by a manufacturer must be audited and the technical file/design dossier (depending on the class of the product) must be reviewed by the notified body selected by the manufacturer. Once the review of the Technical File and audit of the QMS is successful, the Notified Body will issue the CE Marking Certification. For Class I Non-Sterile, Non-Measuring Medical Devices, these devices are not issued with the Notified Body CE marking Certificate as conformity is based on self-certification. For all Class 1 Medical Devices, the device must be registered with the Competent Authority where the EC Rep is located. Most European countries do not require Class IIa, IIb, and III Medical Devices to be registered as it is already subjected to an annual audit by the Notified Body. However, Italy requires all Classes of Medical Devices to be registered.

3.1.7. Step 7: Declaration of Conformity

The Declaration of Conformity is a legally binding document prepared by the manufacturer stating that the device is in compliance with the applicable Directive. It is usually drawn up on a company letterhead and states that the company is in compliance with the relevant regulation and related ISO standards. The declaration of conformity shall also include the following information:

- Identification of the medical device;
- Name of the EC representative;
- Classification of the device;
- Process of compliance;
- Notified body details and CE marking certificate number.

The declaration of conformity should be signed by the company once compliance to EU Regulation can be shown.

3.1.8. Step 8: Labeling and Documentation

There are 31 European countries and 23 official languages, which can make labeling and documentation requirements a bit of a challenge for manufacturers. Although it is not explicitly addressed in the MDD 93/42/EEC, Manufacturers must ensure that the labeling and documentation comply with the National Laws of the European country where the Medical Devices will be sold except if an exemption has been obtained from the Competent Authority. Labeling and documentation must be done in the official language of that state, and manufacturers must ensure that translation into the national language is in compliance with the national regulatory requirements.

3.1.9. Step 9: Monitor Safety and Efficacy

After CE marking has been obtained and devices are now placed on the EU market, Manufacturers must continue to monitor the safety and efficacy of the device. This is usually done by appropriately handling incident reports according to the EU requirements defined in the regulations. Manufacturers must establish an effective marketing surveillance system and review this periodically. Manufacturers are also expected to implement corrective actions as needed.

Chapter

4

History of Medical Devices

CONTENTS

4.1. THE HISTORY OF MEDICAL DEVICE REGULATIONS

Prior to the 1930s, the medical device industry was practically non-existent. Much of the medical devices being manufactured were sold directly to the public without any form of regulations to guide performance, safety, or functionality.

After World War II, the medical device industry and related technologies grew exponentially, leading the first developments of any kind of regulating between industry, government, and the public. By the late 1960s, thousands of medical device manufacturers began to ship their devices. In fact, it was estimated that the total shipped products amounted to a value exceeding $1 billion. In the early 1970s, a released government report documented more than 10,000 injuries as a result of medical devices.

Medical devices have been around for centuries. In fact, evidence dates far back as 7000 BC to the use of scalpels, slings, and splints, which, in essence, are medical devices used in health care.

As modern technology and the application of medical devices merged, so their applications took off in the last 50 years; thereby making medical devices an essential part of health care and patient safety. The key landmarks and trends in medical device development are highlighted in Figure 4.1.

The 1980s saw a surge in the number of patient care medical devices with particular focus on high-resolution imaging devices. In addition, systems for continuous monitoring of cardiovascular parameters such as heart rate and blood pressure were becoming increasingly more standardized in hospitals and health care facilities.

In addition, many health care treatments were being taken over by technological advances in medical devices such as ventilators, kidney dialysis machines, and neonatal incubators. The 1980s to 2000s saw more hospitals in industrialized countries adopting CT scanners and magnetic resonance imaging (MRI) units. The choice of medical devices that could be used grew drastically. Between 2000s and 2010, robotics became a reality within the medical device industry. Assistive devices that could be used with people that had functional disabilities increased. In addition, the use of medical devices integrated with information systems or web-based systems became more and more popular.

1800 - 1850

- FIRST MODERN STETHOSCOPES, LARYNGOSCOPES, OPTHALMOSCOPES

1895

- X-RAYS: discovered by German physicist WIlhelm Roentgen

1903

- FIRST ELECTROCARDIOGRAPH: developed by Dutch physician and physiologist Willem Einthoven. He was awarded the Nobel Prize in 1924 for his discovery

1927

- FIRST MODERN RESPIRATOR: Devised by the United States medical researcher Philip Drinker and his collegues at Harvard University

1928

- FIRST CARDIAC CATHETERIZATION: performed by Werner Forssmann on himself to show feasibility of technique for injecting drugs directly into the heart
- 1940
  - FIRST METALLIC HIP REPLACEMENT SURGERY: performed by United States surgeon Dr. Austin T. Moore

1945

- FIRST KIDNEY DIALYSIS MACHINE: invented by Dutch-born physician Willem Kolff

1950

- FIRST ARTIFICIAL HIP REPLACEMENT (ARTHROPLASTY): Performed by English surgeon Sir John Charnley

1951

- FIRST COMMERCIALLY AVAILABLE ARTIFICIAL HEART VALVE: invented by a United States team led by electrical engineer Miles Edward

1952

- FIRST SUCCESSFUL (EXTRENAL) CARDIAC PACEMAKER: Developed by United States cardiologist Paul Zoll

1960

- FIRST TOTALLY INTERNAL PACEMAKER: Developed by United States electrical engineer Wilson Greatbatch

1970

- FIRST CIMPUTERIZED AXIAL TOMOGRAPHY (CT) SCANNER: Dveveloped by British electrical engineer Godfrey Hounsfield and South Africa born physicist Allen Cormack (jointly awarded the Nobel Prize for medicine in 1979)

1976

- FIRST REGULATORY SYSTEM FOR MEDICAL DEVICES: created by the United States government

1976

- FIRST POSITRON EMISSION TOMOGRAPHY (PET) SCANNING OF PEOPLE: Performed by Abass Alavi, University of Pennsylvania

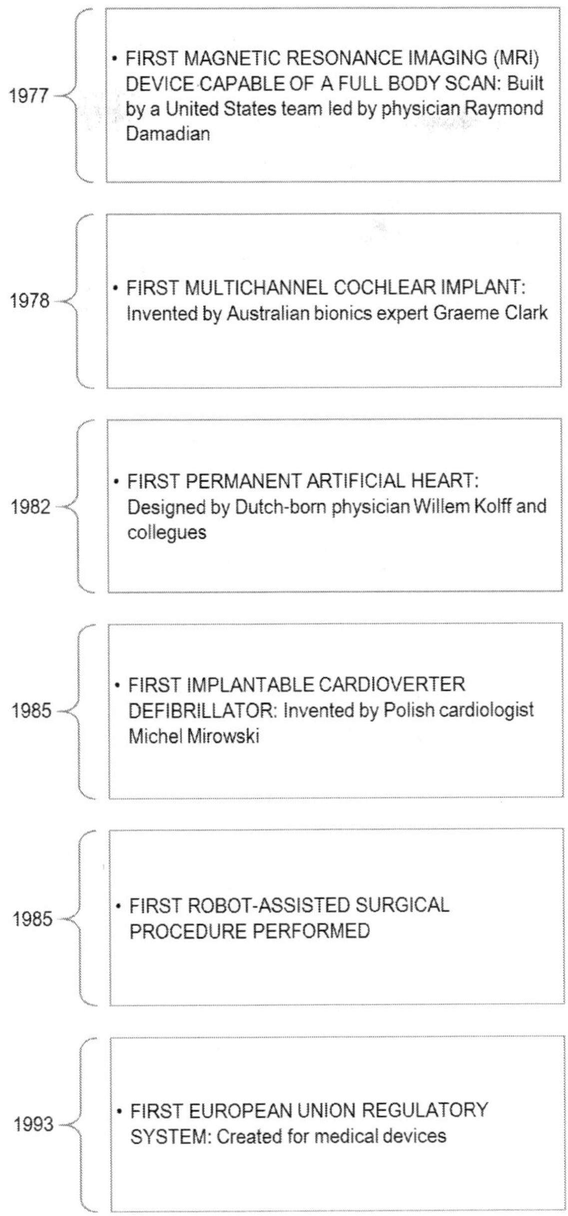

Figure 4.1: Overview of landmarks and key trends in medical device development.

Over the last 20 or so years, the number, range, and complexity of medical devices has increased and as a result, concurrently, so has the complexity of the regulations involved. The last 20 years or so has seen an explosion in the variety, use, and complexity of medical devices. These devices vary greatly in complexity, ranging from implantable devices such as pacemakers to less risky devices such as bedpans and gloves. In 2016 alone, the total global medical device industry was estimated at $130 billion, with almost half of the global production and consumption occurring in the United States alone.

In 2001, the WHO published an "A Model Regulatory Program for Medical Devices," which served as an international guide to assist member states in establishing regulatory programs for medical devices. The aim of this model was to provide information to nations that lacked a medical device regulatory system such that this guide would then enable the production of devices that met internationally compatible regulations. This was then followed by the release of 'Medical Devices Regulations (MDRs). Global overview and guiding principles in 2003 by WHO. The main purpose of this guide was to emphasize the complexity of the medical device industry and identify issues that related to regulations. This document essentially provides guidance to member states who want to create new or modify existing medical device regulatory systems. By 2010, the WHO had then released an estimated report detailing numerically an overview of countries and related medical device regulatory frameworks (Table 4.1).

Table 4.1: Overview of Country Percentages and Related Medical Device Regulatory Frameworks

Countries (%)	Medical Device Regulations
30	Countries that have a developed framework for regulation of medical devices.
30	Countries that partially have regulations of medical devices.
40	Countries that are either developing a framework or do not yet have any regulation.

4.2. WHAT IS A MEDICAL DEVICE?

A medical device is defined as a product or equipment that is intended for use within the health care industry to either diagnose, monitor, or treat diseases or to act as a supportive aid for people with any form of illness or disability. There are over 500,000 different types of medical devices, and these include a vast array of devices such as x-ray machines, infusion pumps and cardiac pacemakers used by doctors, nurses, and surgeons as well as common devices that may be found every day in homes such as wound dressings, thermometers, contact lenses and wound irrigation solutions. According to ISO 13485, medical devices range from simple low-risk products such as urine bags and latex gloves to more complex, technologically advanced programmable devices such as pacemakers. Generally, these devices are grouped into four classes namely Class I, Class IIa, Class IIb and Class III; with Class I being the lowest risk and Class III being the highest risk (Figure 4.2) or Class A, B, C, D with Class A being the lowest risk and Class D being the highest risk.

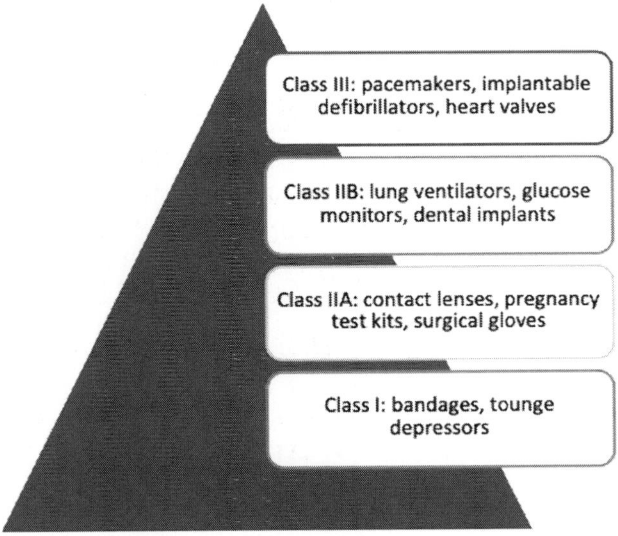

Figure 4.2: Overview of the different medical device classes.

The World Health Organization (WHO) defines a medical device as any instrument, apparatus, implement, machine, appliance, implant, reagent for *in vitro* use, software, material, or other similarly related article intended by the manufacturer to be used, alone or in combination and for human beings in one or more specific medical purpose(s) (Box 4.1). These devices should not achieve its main intended function by pharmacological, immunological, or metabolic means on or in the human body. It may, however, be assisted in its intended function by the abovementioned methods.

Box 4.1: General Uses for Medical Devices

- Diagnosis, prevention, monitoring, treatment, or alleviation of disease;
- Diagnosis, monitoring, treatment, alleviation of, or compensation for an injury;
- Investigation, replacement, modification, or support of the anatomy and/or of a physiological process;
- Supporting or sustaining life;
- Control of conception;
- Disinfection of medical devices;
- Providing information by means of *in vitro* examination of specimens derived from the human body;
- Does not achieve its primary intended action by pharmacological, immunological, or metabolic means, in or on the human body, but may be assisted in its intended function by such means

Source: Derived from: WHO (https://www.who.int/medical_devices/full_deffinition/en/).

The mode of action of a medical device on the human body is what differentiates it from medicines, whose mode of action may often be metabolic, immunological, or pharmacological. In fact, definitions must make clear distinctions on the difference between a medical device and medicine/drug whose primary intended use is obtained through chemical action or by being metabolized in the body. While many regulatory bodies tend to establish their own definitions around this topic, the Global Harmonization Task Force (GHTF) has proposed a harmonized definition for medical devices and *in-vitro* medical devices to ensure uniformity of understanding on a global scale (GHTF document SG1/N029R11) (Box 4.2).

Box 4.2: Definition of a Medical Device as Defined by GHTF (GHTF/SG1/N071:2012)

"Medical device" means any instrument, apparatus, implement, machine, appliance, implant, *in vitro* reagent or calibrator, software, material or other similar or related article, intended by the manufacturer to be used, alone or in combination, for human beings for one or more of the specific purposes of:

- diagnosis, prevention, monitoring, treatment, or alleviation of disease;
- diagnosis, monitoring, treatment, alleviation of or compensation for an injury;
- investigation, replacement, modification, or support of the anatomy or of a physiological process;

- supporting or sustaining life;
- control of conception;
- disinfection of medical devices;
- providing information for medical purposes by means of *in vitro* examination of specimens derived from the human body; and
- which does not achieve its primary intended action in or on the human body by pharmacological, immunological, or metabolic means, but which may be assisted in its function by such means.

The definition laid out by the GHTF covers a multitude of different medical devices, some of which are complex and reflect the latest advances in technological developments such as imaging equipment and implants; as well as those simple devices such as tongue depressors, thermometers, scales, and latex gloves. The GHTF document further defines those produces which may be considered to be medical devices in some jurisdictions while not classified in others. These include: (a) disinfection substances; (b) aids for persons with disabilities; (c) devices incorporating animal and/or human tissues; and (d) devices for *in-vitro* fertilization or assisted reproduction technologies (Box 4.3).

Box 4.3: Definition of an *In-Vitro* Medical Device as Defined by GHTF (GHTF/SG1/N071:2012)

'*In vitro* diagnostic (IVD) medical device' means a medical device, whether used alone or in combination, intended by the manufacturer for the *in-vitro* examination of specimens derived from the human body solely or principally to provide information for diagnostic, monitoring, or compatibility purposes.

4.3. INCIDENTS THAT INFLUENCED CHANGES TO THE REGULATORY FIELD FOR MEDICAL DEVICES

All around the world, medical devices have been involved in various negative incidents on patients which have either injured and, in worse cases, led to patient death. On the other hand, health authorities and various regulatory bodies across the globe have subsequently failed to protect such patients from these poorly functioning (and tested) devices, which have had various negative effects on patients. These include the puncturing of organs, erroneous shocks, rotting of bones and organs, blood poisoning and other effects. A year-long investigation by the International Consortium of Investigative Journalists (ICIJ) in 2018 revealed that many governments actually hold many of these implants (regardless of the complexity or risk classification) to much lower safety testing standards as when compared to pharmaceuticals. Flaws within the regulatory field itself have resulted in medical device manufacturers leaving such problematic devices on the market in some countries while pulling them off the market in other countries. In fact, the Implant Files Investigation conducted by ICIJ which formed one of the biggest and first-ever global examination of the medical device industry revealed just how health and regulatory authorities all around the world have failed to protect millions of patients due to them being uninformed of the crucial risks around such devices. Many of these reported incidents and various past ones have highlighted the need for improvement across standards, processes, procedures, and overall legislation where medical devices are concerned. Many of these incidents have directly highlighted the need for strengthening existing legislation, specifically within the European Union, which has, in turn, led to drastic improvements and changes over the past decade, ultimately giving regulators more control and oversight.

Previous European legislation adopted in the 1990s aimed at ensuring the smooth functioning of the internal market together with an attempt to maintain high health and safety levels. There have however been several key moments in industry that have highlighted the shortcomings of such legislative frameworks.

In April 2017, the new EU MDRs and IVD regulations were adopted by the European Parliament. These new regulations entered into force on the 25 May 2017 and stated that the fundamental revision of the legislation was to provide a 'sustainable regulatory framework for medical devices which ensure a high level of safety and health while supporting innovation.' The main purpose of the changed regulations was to ensure smooth functioning of the internal market with regards to medical devices and enforce a high level of health protection for patients and users.

4.4. PIP BREAST IMPLANT SCANDAL

The PIP breast implant scandal was one of the biggest medical device industry scandals that ultimately influenced and resulted in major regulatory updates, not only within the EU but on a global level. Medical devices in the European Union are regulated by the Medical Devices Directive, which had also been transposed in France in several provisions of the French Public Health Code and in England in the MDRs (2002).

In 1965, plastic surgeon Henri Arion introduced breast implants to France. Together with Jean-Claude Mas, they launched a Poly Implant

Prothese (PIP) company in 1991. This company essentially produced a figure or around 2 million sets of silicone breast implants over a 20-year period, however despite their success, they also created one of the biggest global health scares. This changed the medical device regulatory field.

The health scare came about when PIP decided to change the material being used to manufacture the silicone breast implants to a cheaper, industrial-grade silicone that was not approved for medical use. Industrial grade silicone is thought to have more contaminants than medical-grade silicone; however, it is thought the company chose to use the alternative to save costs and boost overall profits. Investigations later revealed that this industrial grade silicone had a rupture rate that was double that of the general industry average. In addition, later studies revealed that if compromised, the silicone gel was thought to cause both short, and long-term effects such as inflammation and possible scarring in affected patients.

In December 2011, French authorities issued a worldwide alert, advising all women who had been fitted with potentially defective implants to have them removed. By this time, PIP had become the world's third-largest supplier of implants, with over 300,000 women in 65 different countries around the world having received the implants, be they for cosmetic reasons or medical reasons such as reconstructive surgery after breast cancer treatment. Much of the implants had been used widely across Europe. It is estimated that around 40,000 women in Britain and South America alone may have received such implants. Much of the advice given to women from regulatory bodies and surgeons alike varied from country to country. In France, women who had received such implants were urged to have them removed as a precaution. In Venezuela, regulatory authorities further recommended removal while the British government, on the other hand, said there was no evidence to recommend routine removal of PIP implants. They did however advise women who were concerned to have them removed. In Germany, medical groups advised women to seek removal of their implants with a further statement indicating that there was no immediate danger.

The French government subsequently offered to pay for the removal and replacement of all PIP implants fitted in France, while Venezuela said it would cover the cost of implant removals but not the replacements. Within the United Kingdom, those women who had their implants paid for by the National Health Service (NHS) was able to have them removed free of charge after a consultation with their doctor, while private companies who had fitted PIP implants for cosmetic purposes were advised to follow suit.

4.4.1. Overview of Events in the PIP Scandal (Table 4.2)

Table 4.2: Overview of the Events in the PIP Scandal

1991	PIP was launched by Jean-Claude Mas
1992	The United States Food and Drug Administration (FDA) calls for a voluntary moratorium on the use of silicone gel Implants. Implant sales are halted in the United States.
1994	PIP begins selling hydrogel implants.
1996	PIP begins selling saline-filled implants in the United States.
1997	PIP authorized to produce medical-grade silicone implants.
2000	The FDA refuses to approve PIP's saline-filled implants and warns about deviations from good manufacturing practices (GMP) found at the PIP plant. The company withdraws its hydrogel implants from the market as it cannot show they are safe.
2001	PIP starts using an unapproved industrial-grade silicone in their implants.
2003	Regulatory bodies begin to trace the first signs of legal problems and financial losses.
2009	Surgeons started reporting abnormally high rupture rates in France, sparking concern. This led to a host of legal complaints and PIP undergoing bankruptcy. The UK regulatory authority subsequently warned of several medical claims being made.
2010	PIP was placed into liquidation after the French medical safety agency recalled its implants.
2011	The French government recommended that approximately 30,000 women with PIP implants seek removal of the implants as a health precaution.
2012	Jean-Claude Mas, the founder of PIP is arrested. France launches an investigation into the scandal.

2013	Jean-Claude Mas is sent to prison for 4 years and fined 75,000 euros. He is subsequently banned for life from working in medical services or running a company. He also faced two additional legal cases: one for involuntary manslaughter and another linked to the financial implications of the scandal.
2016	Mas's sentence upheld by the appeal court.
2017	New European medical devices regulation (MDR) published and in force, with transition period for manufacturers to comply over 3 years.

4.4.2. What Happened to PIP?

PIP implants were banned, and eventually the company went bankrupt in 2010. The founder of PIP, Jean Claude Mas went under investigation by the French police and was subsequently charged with involuntary injury over the implants. PIP's lawyer eventually did respond to the scandal stating that industrial silicone was used in their implants and that while no tests conducted showed any danger, the main issue was public fear which subsequently created intense media pressure and pushed the blame ultimately onto PIP. Regardless, the concerns raised over the PIP implants have led to one of the biggest calls for changes to regulations within the plastic surgery industry. Ultimately the consumer does not know what should be regarded as being safe or not hence the need for more stringent regulations.

The events of the PIP scandal resulted in the European Commission seriously reviewing and changing their medical device regulations. In fact, the regulation of the medical device industry was already in the process of being reformed when the PIP scandal occurred. The scandal however increased the focus and need for such change. In 2008, the European Commission published a public consultation document that asked stakeholders for their view around the revision of the legal framework of medical devices. This was followed by a public consultation in 2010, which focused on the technical aspects related to the revision of Directive 98/79/EC on IVD medical devices.

4.4.3. Changes to the Medical Device Regulations as a Result of PIP

Ultimately the main reasons for such an industry overall were that, at the time, existing EU regulations, which dated back to the 1990's had not kept pace with the technological progression over the years. This had resulted in many regulatory gaps and uncertainties around products being manufactured from non-viable human tissues or cells as well as implantable and/or other invasive products used for cosmetic purposes. furthermore, there were substantial divergences in the overall interpretation and application of the regulations/rules that emerged between EU countries. Much of this varied interpretation arose from each country having to interpret and respond to such technological and scientific progression individually. In addition, there was limited ability to trace medical devices back to their suppliers.

Essentially the new regulations, now referred to as the European MDR which also includes regulations (EU) 2017/745 and 2017/746 have become far more stringent. Some of the major changes include:

- Expansion of the overall scope within the medical device regulation to cosmetic and/or esthetic devices.
- Clarification regarding areas such as medical software and generic tests.
- Reflection of technological advances through the adaptation of safety and performance requirements that are applicable to new health technologies.
- Stricter requirements for medical device compliance from manufacturers to ensure overall patient safety.
- Stronger supervision of independent assessment bodies by national authorities.
- More power and obligations for those abovementioned assessment bodies to ensure thorough testing and regular checks on manufacturers such as unannounced factory inspections, sample testing.
- Improvement of co-ordination between national surveillance bodies.
- The creation of a national registry to increase traceability. This allows manufacturers to register themselves and the devices they place on the EU market within a central database through a unique device identification system. This way it will make it easier to trace and identify devices and enhance overall post-market safety.
- A ban on aggressive marketing of medical devices.
- Overall requirements for improved training and qualification standards.

- Increased and improved transparency along the entire supply chain of medical devices.
- The establishment of a portal within the European Union, which requires manufacturers to report serious events and corrective actions to reduce the overall risk of such events happening again.
- An increase in the rights and responsibilities of NB as well as the stringency with which they monitor all medical devices.
- A further requirement that within a manufacturers organization to have a qualified person responsible for regulatory compliance.
- Patients that have been implanted with a device should also be given essential information which will allow the device to be identified. This information should also contain necessary warnings or precautions that must be taken.

4.5. THE DEPUY HIP REPLACEMENT RECALL

MoM hip designs were first introduced in the 1950s and 1960s until such time that improvements in manufacturing and measurement technologies sparked designers to revisit the overall design of MoM hips in the 1990s. In 2003, the medical implant manufacturer DuPuy Orthopedics which is a subsidiary of Johnson & Johnson introduced their own version of a metal-on-metal (MoM) hip resurfacing articular surface replacement (ASR) ASR™ and ASR™ XL, a total hip replacement version. Both versions were available in Europe and were available under the clause of 'substantial equivalence' which, allows for fast-track market accreditation through claiming the implant in question is similar to ones already on the market. As per regulatory bodies, the ASR™ hip resurfacing was granted market approval by the FDA in the USA while the ASR™ XL was available globally.

In 2008, problems with ASR™ were first raised at various orthopedic conferences as well as in journal publications. This was followed by an annual report published by the National Joint Registry (NJR) of England and Wales in 2009, which showed high revision rates of approximately 7.5% at 3 years for the ASR™ hip resurfacing. At the time, however, official notices pertaining to such problems were limited to guidance on positioning during surgery until around August 2010 when DuPuy Orthopedics issued a global recall of all ASR™ and ASR™ XL hip systems. In addition, the Medicines and Healthcare products Regulatory Agency (MHRA) further issued a general medical device alert (MDA) on 22 April 2010 for all metal-on-metal (MoM) hip replacements (MDA/2010/033). This MDA did not specifically address the ASR™ implants but rather showed that much data had been available for a while, indicating that general concerns around MoM hip implants did exist. It should also be noted that prior to this, high revision rates with ASR™ were also highlighted in Australia somewhere in 2007. As a consequence of this, the ASR™ was withdrawn at the end of 2009.

4.5.1. Overview of Events in the DuPuy Scandal (Table 4.3)

Table 4.3: Overview of Events in the DuPuy Scandal

2003	ASR™ resurfacing introduced
2007	High revision rates reported by the Australian Orthopedic Association, National Joint Replacement Registry
2008	Subsequent reporting of high revision rates by the Australian Orthopedic Association, National Joint Replacement Registry
2009	First report of high revision rates in National Joint Registry (NJR) of England and Wales. ASR™ withdrawn in Australia and New Zealand
/2010	MHRA issues MDA on all MoM implants noting a small number of patients have adverse reactions; DePuy release guidance on positioning; MHRA issue MDA on positioning; DePuy withdraw the ASR™ globally.

The impact of such high revision rates on patients is significant and should be noted. Failed MoM hip implants can cause both local and systemic health problems as well as symptoms such as hearing, loss, dizziness, decline in cognitive functions, cardiomyopathy, and organ failure while ultimately could result in increased pain and a decrease in overall mobility. Various reasons were attributed to the failure of the hip replacement system. These include loosening and misalignment of components, infection, fracture of the bone, dislocation, metal sensitivity and overall pain. In addition, further complications could include an increase in the level of metal ions in the blood, staining of the bone, necrosis, swelling, and unknown damage to the nerves, tissues, and/or muscles.

4.5.2. Reactions to Published Data on DuPuy Hip Replacement Scandal

In 2007, the Australian Orthopedic Association national joint replacement registry noted that ASR seemed to have twice the risk of revision, thus indicating the need for further surgery as when compared with other resurfacing procedures. This was followed by the registry publishing data the next year showing that some women undergoing MoM hip resurfacing were facing more follow-up operations.

At the beginning of 2010, DuPuy Orthopedics issued a statement saying they were phasing out the ASR Hip Implant due to declining sales; however they failed to mention the data from the Australian implant registry. Later, sometime in March 2010, DuPuy issued their first warning to doctors and patients indicating the high early failure rate but still had not issued a product recall. Eventually, DuPuy Orthopedics withdrew their implants from the market after the NJR in England and Wales shared data that showed patients with their MoM hips having a higher expected revision rate. In fact, data showed that within a 5-year period of receiving DuPuy's resurfacing device, 12% of patients had undergone a revision surgery, and within 5 years of having an ASR total hip replacement, 13% of patients had undergone revision surgery.

It was estimated that about 93 000 people globally had received a DuPuy implant. The first lawsuit against DuPuy Orthopedics was filed in the United States on 15 June, 2010, followed by another on 27 September, 2011. Both lawsuits claimed that DePuy ASR hip replacements were improperly designed and that DePuy knew there were problems with their implants early on and, despite this, did not alert patients or surgeons about the possible problems. In 2014, Johnson & Johnson announced that it would not withdraw a $2.5 billion global settlement. Settlements began at a base of $250,000. Johnson & Johnson continued to deny culpability. It should be further noted that of the total number of people known to receive the implants, approximately 4,700 were thought to be from India. Johnson & Johnson then undertook to working with the Indian government in 2018 in an attempt to support all Indian patients affected by the implants.

4.6. INADEQUATE VAGINAL MESH

Back in the late 1980s, the medical industry was searching for new ways to treat women that were experiencing urinary incontinence as well as vaginal prolapse. Both are common conditions that women experience after childbirth and while most doctors, at the time, suggested physiotherapy, weight-loss, and other non-surgical interventions as a solution; the use of mesh implants seemed like a simple, quick, and convenient alternative. The mesh was initially designed to allow a patient's body tissues to grow through it, making it harder to remove. Mesh implants took about an hour to implant and allowed women to leave the hospital quickly and soon became the standard treatment for millions of women all around the world. While many cases have had effective results, there are also some women who have experienced severe complications such as eroding mesh that has pierced the vaginal wall or bladder, nerve damage, and infections. In addition, many of these women have experienced chronic pain and, in some cases, were barely able to walk.

One of the largest manufacturers of gynecological mesh is medical device manufacturer, Boston Scientific. While the company has been successful in the past, they have also attracted close to 48 000 lawsuits against them, all of which claim that the gynecological mesh used can cause extreme pain and injury. Despite such claims, Boston Scientific has continued to fight such allegations with a comeback stating that over 1 million women have been successfully treated and that they have extensively tested the plastic/resin used to confirm the mesh composition, safety, and performance. Unfortunately, many doctors have also found otherwise, with some claiming the mesh causes a chronic inflammatory reaction and stating that once removed, the mesh seems to have a substantially altered architecture. In addition, they seem to have shrunk by 50% in width and are excised in scar tissue.

Investigations revealed that the mesh is made of a polypropylene plastic which is known to be a common packaging material. Boston Scientific however has proof that they have clearance from the FDA to use a particular brand of polypropylene called Marlex. This brand is said to be made in Texas by a subsidiary of Chevron Phillips. In 2004 however, Chevron Phillips became concerned with the use of Marlex in the medical industry and issued a warning that it must not be used for any form of permanent implantation in the body. This is due to the unstable oxidative state of polypropylene. Oxygen is known to break up the polypropylene and cause dissipation of antioxidant additives over time and as a result, causes the polypropylene to fall apart within the body. As time passed, Chevron Phillips cut off Boston Scientific's supply of Marlex, somewhere in 2005; however, Boston Scientific remained adamant on continuing with their processes. As a result, they continued their search for a supplier and eventually came across a broker in China who claimed to have large amounts of Marlex that was supposedly imported from Chevron Phillips in Texas. Simultaneously that same month, the FDA issued a report claiming that over a 5-year period, they had found about 4000 reports of injury, death, and malfunction as well as associated complications including pain, infection, urinary problems, bleeding, and organ perforation. This further gave the industry a reason to believe that if they planned on switching plastics, the FDA would now require years of testing, which ultimately might fail and hinder sales.

The Chinese company seemed like the best option at the time. Boston Scientific had established their own procedures for what required documents would be needed as well as requests for various import records to prove the Marlex was really imported from Texas by the Chinese company. Unfortunately, the Chinese company did not have any documents that could ensure the authenticity of the Marlex could be verified. Further investigation revealed that lot numbers on the bags did not match up as well as counterfeit printing. Boston Scientific then further ordered testing on the material and analyzed 11 different parameters. Results from nine of these tests differed. Despite this, Boston Scientific still bought a large amount of the Chinese plastic and continued to use it in their manufacturing processes.

4.7. THE INFLUENCE OF TECHNOLOGY TRENDS ON MEDICAL DEVICE DEVELOPMENT

As technology advances, so does the development of smaller and less expensive robotic systems that ideally allow high-precision surgeries to continue. The need for synergy will continue to direct future innovation and medical device design while nanotechnology and genomics will interact and continue to improve personalized care. Research and development officers within manufacturing firms may be swayed by the lure of technology; however, ideally, what is needed is a balance between technology, cost-effectiveness, need, and overall usefulness of the device in industry. One of the biggest examples supporting this trend is among the medical imaging facilities whose main aim is to promote whole-body CT scanning as a means for preventative measures among individuals with no none symptoms or suspected disease implications. The US FDA announced in July 2009 that it had not cleared or approved any CT system specifically for use in screening as no manufacturer has ever been able to demonstrate effective disease screening with their CT scanner. In addition, they also made reference to the risk of radiation doses through repetitive CT examinations. As technology changes and improves, so trends develop; all of which have hidden implications on the use of medical devices and their expected health implications (Table 4.4).

Table 4.4: Current Trends Which Have Had or will Continue to Have a Significant Effect on Medical Device Use and Related Implications for Health

Technology Convergence	Medical device implications	• labor-intensive • costly • requires multi-disciplinary teams • highly dependent on infrastructure • difficult to apply in low-resource settings • likely to raise patient expectations • patient pressure on health care • likely to be reserved for financially privileged patients • liable to incur technical risks • patient safety problems may arise
	Health implications	• could enhance sharing of clinical information • could reduce errors of data entry • could facilitate data analysis • could enhance clinician efficiency • could allow increased caseload • could improve patient outcomes • could improve patient safety by giving clinicians access to all patient data
Decentralization of Care Delivery	Medical device implications	• reliance on portable devices • may result in requiring of greater ruggedness than stationary hospital equipment • reliance on non-medical technology such as communication networks, durable battery technology, and power sources
	Health implications	• benefit to patients and family caregivers • adequate user training required • may result in more patients receiving care outside a hospital setting • more efficient patient monitoring • more efficient health care • enables access of rural health-care professionals to highly specialized clinical procedures

Computer-Aided Surgery and Robotics	Medical device implications	• risk of use error • risk of system malfunctioning • may lead to patient injury • longer learning curve for users • requires integration of imaging and surgical navigation systems • device-specific training required • high acquisition costs • high procedure costs • could be prohibitive for many resource-scarce settings • maintenance and quality control requirements likely to be demanding
	Health implications	• greater accuracy of surgical procedures • long initial learning curve • could still save time • greater consistency where reproduction of procedures is concerned • could circumvent obstacles to surgical intervention • allows frequent checking of instrument status and positioning • could give nurses more time with patients by alleviating menial tasks

4.8. THE GLOBAL MEDICAL DEVICE MARKET

The global medical device market is enormous. In 2008, revenue from sales of medical devices worldwide was estimated at little over US$ 210 billion. This is close to double that of the estimated revenue for 2001. Globally, the medical device industry comprises of more than 27,000 medical device companies. Four-fifths of the global medical device sales revenue comes from sales in the United States and Europe, with around 10 countries accounting for approximately 80% of global sales revenue (Figure 4.3).

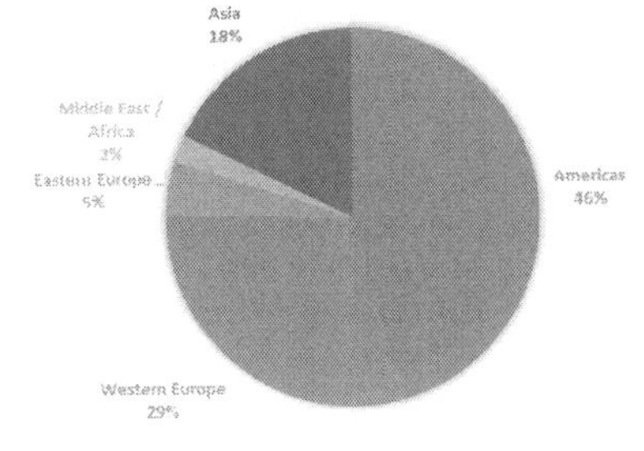

Figure 4.3: Medical device markets by region based on 2009 percent sales revenue.

Source: Ref.: The World Medical Markets Fact Book (2009).

The United States is at the top of that list accounting for around 40% of sales, closely followed by Japan (10%), Germany (8%) and France (4%) (Table 4.5).

Table 4.5: Top 10 Countries by 2009 Sales Revenue

Ranking	Country	Sales Revenue US$ (Millions)	Percentage (%)
1	United States	91,316	40.7%
2	Japan	22,721	10.1%
3	Germany	18,147	8.1%
4	France	8,625	3.8%
5	Italy	8,004	3.8%
6	United Kingdom	7,628	3.4%
7	China	6,161	2.7%
8	Spain	4,887	2.2%
9	Canada	4,757	2.1%
10	Switzerland	4,063	1.8%
Subtotal		**176,309**	**78.7%**
World Total (67 Countries)		**224,103**	**100%**

Source: Ref.: The World Medical Markets Fact Book (2009).

There are about 27,000 medical device companies in the world. Sales revenues indicate that the majority of top 30 medical device companies have their headquarters in the United States of America (Table 4.6). Together, these companies account for 89 and of the estimated global sales revenue of US$ 210 billion. The remaining 11% is shared by a vast number of manufacturers in the small to medical enterprise (SME) category.

Table 4.6: Top 30 Medical Device Companies based on Sales Revenue

Ranking	Company	Headquarters	Sales Revenue US$ (Millions)
1	Johnson & Johnson	United States	23,225
2	GE Healthcare	United States	17,392
3	Siemens Healthcare	Germany	15,526
4	Medtronic	United States	13,515

5	Baxter International	United States	12,400
6	Covidien	Ireland	9,910
7	Philips Healthcare	Netherlands	9,227
8	Boston Scientific	United States	8,050
9	Becton Dickinson	United States	7,156
10	Stryker	United States	6,718
11	B. Braun	Germany	5,263
12	Cardinal Health	Ireland	4,600
13	St. Jude Medical	United States	4,363
14	3M Health Care	United States	4,293
15	Zimmer	United States	4,121
16	Olympus	Japan	3,920
17	Smith & Nephew	United Kingdom	3,801
18	Hospira	United States	3,620
19	Terumo	Japan	3,400
20	Danaher Corporation	United States	3,227
21	Synthes	United States	3,206
22	Beckman Coulter	United States	3,099
23	Alcon	Switzerland	2,881
24	Fresenius Medical Care	Germany	2,875
25	CR. Bard	United States	2,452
26	Abbott	United States	2,241
27	Dentsply	United States	2,194
28	Varian Medical	United States	2,070
29	Biomet	United States	2,135
30	Drager	Germany	1,729
			188,609

It is derived from company annual reports (listed in the references section). It should be noted that not all data refers strictly to medical devices. For those companies that compete in several industrial sectors, there is no differentiation between sales revenues from medical devices and other products.

Historically, however, the majority of high-tech medical devices have been manufactured by companies based in first world/industrialized countries, while low tech devices (such as condoms, surgical gloves, gauze, and so on) have been manufactured in countries with emerging economies such as India, Indonesia, Malaysia, and Sri Lanka.

In addition, many of the multinational companies listed above tend to have manufacturing sites based in developing countries due to the cheaper costs.

4.9. UNDERSTANDING THE DIFFERENCES BETWEEN MEDICINES AND MEDICAL DEVICES

Medicines and medical devices are actually quite similar in various aspects. Both form health technologies which society needs to survive and ensure healthy living, both can be used to diagnose, treat, alleviate, and cure diseases. Both require some form of regulatory control and both need an adequately implemented post-market surveillance system.

In addition, both have some form of intellectual property, need supply chain, and have become crucial to modern health care. Despite the numerous similarities, medical devices differ from medicines in various ways (Table 4.8).

Table 4.8: Key Points in the Differences Between Medical Devices and Medicines

Diversity	**Medical Devices Vary in Size, Complexity, Packaging, and Use**
Innovation	Innovation of medical devices results mainly from the insights of clinicians rather than laboratory studies. Medical devices usually undergo small/incremental improvements. They have a relatively short commercial lifecycle of about 18 months on average.
Durability	Medical devices have a wide range of durability with extremes ranging from a few minutes for disposable devices to several decades for some implantable devices and medical equipment.

Mode of action	Medical devices do not achieve their principal intended action in or on the human body by pharmacological, immunological, or metabolic means, although some devices such as syringes may be used to deliver medicines. Medical devices produce mainly local and physical effects on the body rather than systemic and pharmacological effects.
Regulation	The risk class of the medical device in question will determine the extent of regulatory scrutiny involved. Often the assessment of safety and efficacy for low-risk classes of medical devices can be performed by the manufacturer themselves. For high-risk classes of medical devices, some form of evidence may be submitted to the competent authorities (CA) to prove safety and efficacy. Efficacy or effectiveness of medical devices is proven before they are put on the market, but it should be noted that clinical effectiveness is more difficult to prove.
Supply	About 80% of the medical device industry is made up of small and medium enterprises. Distribution of heavy medical equipment is usually costly. There is no well-defined supply chain or profession involved in the supply of medical devices, which is in direct contrast to that of medicines where pharmacists may aid in the supply.
Usage	The performance of a medical device depends not only on the device itself but also on how it is used. The user interface of a medical device is usually not direct; for example, device-to-patient interaction. In many cases, there is an intermediary interface involving a device-operator-patient interaction. Often medical devices are associated with some form of learning. This is particularly true with more complex high-tech devices which may need a form of technical training and support. Medical devices may require service and maintenance. Many medical devices are used for diagnostic purposes. Many medical devices are used to alleviate functional disabilities, and this must be taken into consideration.

Medical devices are often based on mechanical, biomedical engineering techniques and may either permanently, or temporarily replace a body function while medicines, on the other hand are designed to interact with the body's chemistry. The efficacy of a medicine on the one hand is much simpler to demonstrate than for a medical device as this depends on many more variables such as the setting in which the device will operate, the individual patient characteristics as well as the skill and experience of the health care practitioner using or applying the device. Furthermore, the research and development models of medical devices as compared to medicines are extremely different. Most medical devices cannot be evaluated using a randomized clinical trial as is often the case for medicines. They are, however, evaluated for their overall efficacy and safety based on data acquired from clinical investigations.

While both medical devices and medicines are regulated, the medical device industry has its own unique regulatory system when compared to medicines. Medical devices tend to have shorter product life cycles with technical improvements taking an estimate of 2 years as opposed to decades for medicines. For this reason, in particular, trying to apply the same regulations of medicines to medical devices would thus delay access to vital medical resources and procedures while not really increasing patient safety. Lastly, it should be noted that the sales and marketing models differ for medical devices as when compared to medicines. Where medicines are concerned, it is mainly the prescribing doctor who becomes the target of sales and marketing; on the other hand, for medical devices; there are many more stakeholders concerned who can influence the overall adoption and use of a device. These often include hospital managers, nurses, and so on. for this reason, in particular, medical devices have a greater emphasis on training, education, service, and overall distribution and maintenance; all of which can greatly affect business models (Table 4.9).

Table 4.9: Detailed Differences between Medical Devices and Medicines

Medical Devices	***In Vitro* Diagnostic Medical Devices**	**Medicines**
In vivo and/or *ex vivo* uses	*In vitro* use	*In vivo* use
Diagnostic or therapeutic intended use	Diagnostic intended use	Therapeutic intended use
Outcomes of use often depend directly on skill or experience of the user	Outcomes generally not dependent on skill or experience of user	Outcomes generally not dependent on skill or experience of user

Active components: • Generally based on mechanical, electrical, and materials engineering. • Many medical devices incorporate and are driven by software.	Active components: • IVD components have no therapeutic effect. They are used only for diagnosis. The key components are those that are essential for the detection of the analyte of interest. • Core reagents are biological (for example, antibodies) • Performance of various tests (such as sensitivity and/or specificity) depends on the design of test, geographic variations of the infective agent, populations, and the setting of use. • Variable batch sizes for a given reagent. • Individual batches of the same reagent may use different starting materials. • Stability varies between products and may vary between batches. • Generally stored at 4°–8°C. • Generally short shelf lives (< 12 months)	Active components: • Based on pharmacology and chemistry; now encompassing biotechnology, genetic engineering, and so on. • Pharmacologic properties and action of active ingredients are known. These are based on pre-clinical and clinical studies. • Standardized batch sizes, manufacturing processes and starting materials. • Products are stable. • Products are generally stored at room temperature. • Generally, have long shelf lives.
Product development: • Can be used for a wide variety of products and applications from thermometers and bandages to pacemakers to x-rays. • Designed to perform specific functions and approved on the basis of safety and performance. • Many products developed by doctors or nurse	Product development: • Products are usually in the form of reagents. • Wide variety of IVDs designed for different indications such as screening, confirmation, and monitoring of treatment. • Product development by discovery and evaluation. • Approved on basis of performance. • Products developed in laboratories by chemists and biologists.	Product development: • Products are usually in the form of pills, solutions, aerosols, or ointments. • Product development by discovery and trial. • Approved on basis of safety and efficacy. • Products developed in laboratories by chemists and pharmacologists.

Most act through physical interaction with the body or body part.	Tests performed on samples (such as blood, tissues, saliva, feces, or urine) that have been taken from the body. No direct contact with the human body. No need for clinical trials with patients. Generally, "performance evaluations" of IVDs conducted with samples.	Products are administered by mouth, skin, eyes, inhalation, or injection and are biologically active. Products are effective when absorbed into the human body. Products often act systemically on the entire body
Concerns around intellectual property: • Continuous innovation and iterative improvements based on new science, new technology, and new materials	Concerns around intellectual property: • Continuous innovation and iterative improvements based on new science, new technology, and new materials	Concerns around intellectual property: • Extensive research and development of a specific compound or molecule; takes several years for a new drug to enter the product pipeline
Short product life cycle and investment recovery period (typically 18 months on market). Little patent linkage possible. Data exclusivity is important.	Rapid innovation, short product life cycle (3–5 years) and investment recovery period	Intensive patent protection, including data exclusivity and patent linkage, needed due to extensive product life cycle and long investment recovery period.
The majority of new products bring added functions and clinical value based on incremental improvements; this often results in a range of related "models" for a user to choose among based on patient needs.	Majority of new products bring added clinical value based on improvements of performance	Usually, large step innovation
Support provided: • Large investment in manufacturing, distribution, and training/education (and retraining) • Need to provide service and maintenance (for many high-tech devices)	Support provided: • Large investment in manufacturing, distribution, and training/education (and retraining) • Need to provide service and maintenance (for many high-tech devices)	Support provided: • Low manufacturing and distribution cost, and, in most cases, little or no training, service or maintenance costs.

Chapter

5

The Medical Device Directive 93/42 EEC

CONTENTS

5.1. THE MEDICAL DEVICE DIRECTIVE (MDD)

5.1.1. What Is the Medical Device Directive (MDD)?

Any medical device placed on the market must comply with the relevant legislation of the country in which that device is entering. For Europe, this was the medical device directive (MDD); in transition to the medical device regulation (MDR). This means, for a manufacturer to legally place a medical device on the European market, they must show that their device meets the essential requirements (ER) of the relevant regulation.

The legislation consists of three directives:

- Directive 90/385/EEC for active implantable medical devices (AIMD);
- Directive 93/42/EEC for medical devices (MDD); and
- Directive 98/79/EC for *in vitro* diagnostic medical devices (IVDD).

As such, all devices being sold within the European Union were required to conform to the directive by 1 June, 1998. It is important to understand that the main purpose of the MDD was to harmonize those standards that benefit manufacturers, users, and patients while also defining the specific requirements for clinical testing, design, manufacture, testing/inspection, marketing, installation, and service of medical devices being sold within the European Union.

5.2. UNDERSTANDING THE MDD: LET US TAKE A CLOSER LOOK

The MDD consists of 23 articles, 12 annexes, and 18 classification rules. Within the MDD, three main areas of focus for medical devices were:

- The essential requirements;
- Classification rules; and
- Conformity routes for assessments.

5.2.1. Classification Rules under the MDD

The MDD groups all medical devices into one of four classes, namely Class I, IIa, IIb, and III. Each class is of increasing risk to the patient with medical devices being placed in each respective class based on their safety, function, and intended purpose. Class I devices are those that pose a low risk to the patient and, except for sterile products or measuring devices, can be self-certified by the manufacturer.

These devices do not come into contact or interact with the body. Class IIa devices are of a medium risk and are invasive in their interaction with the human body.

The methods on invasion however are limited to natural body orifices. Class IIb devices are of a medium risk and are either partially or implantable within the human body, and may modify the biological or chemical composition of body fluids. Class III devices are of high risk and require design/clinical trial reviews, product certification and an assessed quality system. These devices affect the functioning of vital organs and/or life-support systems (Figure 5.1).

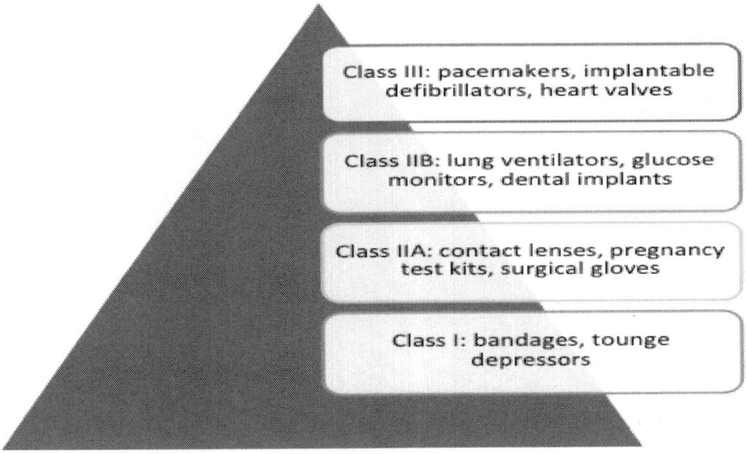

Figure 5.1: Overview of the different medical device classes.

The level of control applied to a medical device is often proportionate to the level of risk it possesses in order to ensure the protection of patient health. Medical device classification may also be affected by the time period in which the device performs its intended function.

Three definitions for duration of use apply to the directive: *transient* (normally intended for continuous use of less than 60 minutes), *short-term* (normally intended for continuous use of 30 days or less) and *long-term* (normally intended for continuous use of more than 30 days).

Article 9 of the MDD 93/42/EEC: Classification
1. Devices shall be divided into Classes I, IIa, IIb, and III. Classification shall be carried out in accordance with Annex IX. 2. In the event of a dispute between the manufacturer and the notified body concerned, resulting from the application of the classification rules, the matter shall be referred for decision to the competent authority to which the notified body is subject. 3. Where a Member State considers that the classification rules set out in Annex IX require adaptation in the light of technical progress and any information which becomes available under the information system provided for in Article 10, it may submit a duly substantiated request to the Commission and ask it to take the necessary measures for adaptation of classification rules. The measures designed to amend non-essential elements of this Directive relating to adaptation of classification rules shall be adopted in accordance with the regulatory procedure with scrutiny referred to in Article 7(3).

Now let us look at the requirements of Annex IX (Classification criteria) in greater detail.

Annex IX of the MDD 93/42/EEC: Classification Criteria
I. Definitions: ***1. Definitions for the Classification Rules:*** *1.1. Duration:* • **Transient:** Normally intended for continuous use for less than 60 minutes. • **Short-Term:** Normally intended for continuous use for not more than 30 days. • **Long-Term:** Normally intended for continuous use for more than 30 days. *1.2. Invasive Devices:* • **Invasive Device:** A device which, in whole or in part, penetrates inside the body, either through a body orifice or through the surface of the body. • **Body Orifice:** Any natural opening in the body, as well as the external surface of the eyeball, or any permanent artificial opening, such as a stoma. • **Surgically Invasive Device:** An invasive device which penetrates inside the body through the surface of the body, with the aid or in the context of a surgical operation. For the purposes of this Directive, devices other than those referred to in the previous subparagraph and which produce penetration other than through an established body orifice shall be treated as surgically invasive devices.

- **Implantable Device:** Any device which is intended:
  - o to be totally introduced into the human body or;
  - o to replace an epithelial surface or the surface of the eye, by surgical intervention which is intended to remain in place after the procedure.

Any device intended to be partially introduced into the human body through surgical intervention and intended to remain in place after the procedure for at least 30 days is also considered an implantable device.

1.3. Reusable Surgical Instrument:

Instrument intended for surgical use by cutting, drilling, sawing, scratching, scraping, clamping, retracting, clipping or similar procedures, without connection to any active medical device and which can be reused after appropriate procedures have been carried out.

1.4. Active Medical Device:

Any medical device operation of which depends on a source of electrical energy or any source of power other than that directly generated by the human body or gravity and which acts by converting this energy. Medical devices intended to transmit energy, substances or other elements between an active medical device and the patient, without any significant change, are not considered to be active medical devices. Standalone software is considered to be an active medical device.

1.5. Active Therapeutical Device:

Any active medical device, whether used alone or in combination with other medical devices, to support, modify, replace, or restore biological functions or structures with a view to treatment or alleviation of an illness, injury, or handicap.

1.6. Active Device for Diagnosis:

Any active medical device, whether used alone or in combination with other medical devices, to supply information for detecting, diagnosing, monitoring, or treating physiological conditions, states of health, illnesses, or congenital deformities.

1.7. Central Circulatory System:

For the purposes of this Directive, 'central circulatory system' means the following vessels:

arteriae pulmonales, aorta ascendens, arcus aorta, aorta descendens to thebifurcatio aortae, arteriae coronariae, arteria carotiscommunis, arteriacarotis externa, arteria carotisinterna, arteriae cerebrales, truncus brachiocephalicus, venae cordis, venae pulmonales, vena cava superior, vena cavainferior.

1.8. Central Nervous System:

For the purposes of this Directive, 'central nervous system' means brain, meninges, and spinal cord.

II. Implementing Rules:

2. Implementing Rules:

2.1. Application of the classification rules shall be governed by the intended purpose of the devices.

2.2. If the device is intended to be used in combination with another device, the classification rules shall apply separately to each of the devices. Accessories are classified in their own right separately from the device with which they are used.

2.3. Software, which drives a device or influences the use of a device, falls automatically in the same class.

2.4. If the device is not intended to be used solely or principally in a specific part of the body, it must be considered and classified on the basis of the most critical specified use.

2.5. If several rules apply to the same device, based on the performance specified for the device by the manufacturer, the strictest rules resulting in the higher classification shall apply.

2.6. In calculating the duration referred to in Section 1.1 of Chapter I, continuous use means 'an uninterrupted actual use of the device for the intended purpose.' However, where usage of a device is discontinued in order for the device to be replaced immediately by the same or an identical device, this shall be considered an extension of the continuous use of the device.

III. Classification:

1. Non-Invasive Devices:

1.1. Rule 1:

All non-invasive devices are in Class I, unless one of the rules set out hereinafter applies.

1.2. Rule 2:

All non-invasive devices intended for channeling or storing blood, body-liquids or tissues, liquids, or gasses for the purpose of eventual infusion, administration or introduction into the body are in Class IIa:

- If they may be connected to an active medical device in Class IIa or a higher class.
- If they are intended for use for storing or channeling blood or other body liquids or for storing organs, parts of organs or body tissues, in all other cases they are in Class I.

1.3. Rule 3:

All non-invasive devices intended for modifying the biological or chemical composition of blood, other body liquids or other liquids intended for infusion into the body are in Class IIb, unless the treatment consists of filtration, centrifugation or exchanges of gas, heat, in which case they are in Class IIa.

1.4. Rule 4:

All non-invasive devices which come into contact with injured skin:

- Are in Class I if they are intended to be used as a mechanical barrier, for compression or for absorption of exudates.

• Are in Class IIb if they are intended to be used principally with wounds which have breached the dermis and can only heal by secondary intent.

• Are in Class IIa in all other cases, including devices principally intended to manage the micro-environment of a wound.

2. Invasive Devices:

2.1. Rule 5:

All invasive devices with respect to body orifices, other than surgically invasive devices and which are not intended for connection to an active medical device or which are intended for connection to an active medical device in Class I:

• Are in Class I if they are intended for transient use.

• Are in Class IIa if they are intended for short-term use, except if they are used in the oral cavity as far as the pharynx, in an ear canal up to the eardrum or in a nasal cavity, in which case they are in Class I.

• Are in Class IIb if they are intended for long-term use, except if they are used in the oral cavity as far as the pharynx, in an ear canal up to the eardrum or in a nasal cavity and are not liable to be absorbed by the mucous membrane, in which case they are in Class IIa.

All invasive devices with respect to body orifices, other than surgically invasive devices, intended for connection to an active medical device in Class IIa or a higher class, are in Class IIa.

2.2. Rule 6:

All surgically invasive devices intended for transient use are in Class IIa unless they are:

- Intended specifically to control, diagnose, monitor, or correct a defect of the heart or of the central circulatory system through direct contact with these parts of the body, in which case they are in Class III.
- Reusable surgical instruments, in which case they are in Class I.
- Intended specifically for use in direct contact with the central nervous system, in which case they are in Class III.
- Intended to supply energy in the form of ionizing radiation, in which case they are in Class IIb.
- Intended to have a biological effect or to be wholly or mainly absorbed in which case they are in Class IIb.
- Intended to administer medicines by means of a delivery system, if this is done in a manner that is potentially hazardous taking account of the mode of application, in which case they are in Class IIb.

2.3. Rule 7:

All surgically invasive devices intended for short-term use are in Class IIa unless they are intended:

- Either specifically to control, diagnose, monitor, or correct a defect of the heart or of the central circulatory system through direct contact with these parts of the body, in which case they are in Class III.

- Or specifically for use in direct contact with the central nervous system, in which case they are in Class III.
- Or to supply energy in the form of ionizing radiation, in which case they are in Class IIb.
- Or to have a biological effect or to be wholly or mainly absorbed in which case they are in Class III.
- Or to undergo chemical change in the body, except if the devices are placed in the teeth, or to administer medicines, in which case they are in Class IIb.

2.4. Rule 8:

All implantable devices and long-term surgically invasive devices are in Class IIb unless they are intended:

- To be placed in the teeth, in which case they are in Class IIa.
- To be used in direct contact with the heart, the central circulatory system, or the central nervous system, in which case they are in Class III.
- To have a biological effect or to be wholly or mainly absorbed, in which case they are in Class III.
- Or to undergo chemical change in the body, except if the devices are placed in the teeth, or to administer medicines, in which case they are in Class III.

3. Additional Rules Applicable to Active Devices:

3.1. Rule 9:

All active therapeutic devices intended to administer or exchange energy are in Class IIa unless their characteristics are such that they may administer or exchange energy to or from the human body in a potentially hazardous way, taking account of the nature, the density and site of application of the energy, in which case they are in Class IIb. All active devices intended to control or monitor the performance of active therapeutic devices in Class IIb, or intended directly to influence the performance of such devices are in Class IIb.

3.2. Rule 10:

Active devices intended for diagnosis are in Class IIa:

- If they are intended to supply energy which will be absorbed by the human body, except for devices used to illuminate the patient's body, in the visible spectrum.
- If they are intended to image *in vivo* distribution of radiopharmaceuticals.
- If they are intended to allow direct diagnosis or monitoring of vital physiological processes, unless they are specifically intended for monitoring of vital physiological parameters, where the nature of variations is such that it could result in immediate danger to the patient, for instance variations in cardiac performance, respiration, activity of CNS in which case they are in Class IIb.

Active devices intended to emit ionizing radiation and intended for diagnostic and therapeutic interventional radiology, including devices which control or monitor such devices, or which directly influence their performance, are in Class IIb.

3.3 Rule 11:
All active devices intended to administer and/or remove medicines, body-liquids or other substances to or from the body are in Class IIa, unless this is done in a manner:

- That is potentially hazardous, taking account of the nature of the substances involved, of the part of the body concerned and of the mode of application in which case they are in Class IIb.

3.4. Rule 12:
All other active devices are in Class I.

4. Special Rules:

4.1. Rule 13:
All devices incorporating, as an integral part, a substance which, if used separately, can be considered to be a medicinal product, as defined in Article 1 of Directive 2001/83/EC, and which is liable to act on the human body with action ancillary to that of the devices, are in Class III.
All devices incorporating, as an integral part, a human blood derivative is in Class III.

4.2. Rule 14:
All devices used for contraception or the prevention of the transmission of sexually transmitted diseases are in Class IIb, unless they are implantable or long-term invasive devices, in which case they are in Class III.

4.3. Rule 15:
All devices intended specifically to be used for disinfecting, cleaning, rinsing, or, when appropriate, hydrating contact lenses are in Class IIb.
All devices intended specifically to be used for disinfecting medical devices are in Class IIa. Unless they are specifically to be used for disinfecting invasive devices, in which case, they are in Class IIb.
This rule does not apply to products that are intended to clean medical devices other than contact lenses by means of physical action.

4.4. Rule 16:
Devices specifically intended for recording of X-ray diagnostic images are in Class IIa.

4.5. Rule 17:
All devices manufactured utilizing animal tissues or derivatives rendered non-viable are Class III except where such devices are intended to come into contact with intact skin only.

*4.6. R*ule 18:
By derogation from other rules, blood bags are in Class IIb.

5.2.2. How to Comply with the Medical Device Directive (MDD) 93/42/EEC?

In summary, showing compliance with the MDD can be summarized as follows:

Before proceeding with an assessment procedure, it is important to establish whether the manufacturer, can assess the product by themselves or whether there is a need to involve a Notified Body. In order to do that, the manufacturer must first determine under which class the medical devices fall. The involvement of a Notified Body is not necessary for medical devices of Class I unless they have a measuring function or are placed on the market in a sterile condition. For all medical devices belonging to Class III, and for medical devices belonging to Class IIa and IIb on a representative basis, the design of the medical device and its compliance with the ER must be examined by a Notified Body. The Notified Body issues then, a certificate that indicates, by referring to one of the Annexes II to VI of the MDD, what has been verified. In the case of devices falling within Class I (low risk), other than devices which are custom-made or intended for clinical investigations, the manufacturer performs the conformity assessment and documents the assessment on their own. In this case, the manufacturer shall follow the procedure referred to in Annex VII and draw up the EC declaration of conformity required before placing the device on the market.

In the case of devices falling within Class IIa (medium risk), other than devices which are custom-made or intended for clinical investigations, the manufacturer performs the conformity assessment and documents the assessment in his own right. And, in order to affix the CE marking, shall follow the procedure relating to the EC declaration of conformity set out in Annex VII, coupled with either:

- The procedure relating to the EC verification set out in Annex IV;
- The procedure relating to the EC declaration of conformity set out in Annex V (production quality assurance); or
- The procedure relating to the EC declaration of conformity set out in Annex VI (product quality assurance).

Instead of applying these procedures, the manufacturer may also follow the procedure relating to the EC declaration of conformity set out in Annex II (full quality assurance).

In the case of devices falling within Class IIb (medium risk), other than devices which are custom-made or intended for clinical investigations, the manufacturer shall, in order to affix the CE marking, either:

i. Follow the procedure relating to the EC declaration of conformity set out in Annex II (full quality assurance); in this case, point 4 of Annex II is not applicable; or

ii. Follow the procedure relating to the EC type-examination set out in Annex III, coupled with:

a. The procedure relating to the EC verification set out in Annex IV;

b. The procedure relating to the EC declaration of conformity set out in Annex V (production quality assurance); or

c. The procedure relating to the EC declaration of conformity set out in Annex VI (product quality assurance).

In the case of devices falling within Class III (high risk), other than devices which are custom-made or intended for clinical investigations, the manufacturer shall, in order to affix the CE marking, either:

i. Follow the procedure relating to the EC declaration of conformity set out in Annex II (full quality assurance); or

ii. Follow the procedure relating to the EC type-examination set out in Annex III, coupled with:

a. The procedure relating to the EC verification set out in Annex IV; or

b. The procedure relating to the EC declaration of conformity set out in Annex V (production quality assurance).

5.2.3. Understanding Which Products Are Within the Scope of the Medical Device Directive (MDD) 93/42/EEC

Article 1 of the MDD defines a 'medical device' to mean any instrument, apparatus, appliance, software, material, or other article, whether used alone or in combination, including the software intended by its manufacturer to be used specifically for diagnostic and/or therapeutic purposes and necessary for its proper application, intended by the manufacturer to be used for human beings for the purpose of:

- Diagnosis, prevention, monitoring, treatment, or alleviation of disease;
- Diagnosis, monitoring, treatment, alleviation of or compensation for an injury or handicap;
- Investigation, replacement, or modification of the anatomy or of a physiological process;
- Control of conception; and which does not achieve its principal intended action in or on the human body by pharmacological, immunological, or metabolic means, but which may be assisted in its function by such means;

In addition, the scope includes an 'accessory,' which means an article which, whilst not being a device is intended specifically by its manufacturer to be used together with a device to enable it to be used.

Article 1 of the MDD 93/42/EEC: Definitions, Scope

1. This Directive shall apply to medical devices and their accessories. For the purposes of this Directive, accessories shall be treated as medical devices in their own right. Both medical devices and accessories shall hereinafter be termed devices.
2. For the purposes of this Directive, the following definitions shall apply:

a. 'medical device' means any instrument, apparatus, appliance, software, material, or other article, whether used alone or in combination, including the software intended by its manufacturer to be used specifically for diagnostic and/or therapeutic purposes and necessary for its proper application, intended by the manufacturer to be used for human beings for the purpose of:

- Diagnosis, prevention, monitoring, treatment, or alleviation of disease.
- Diagnosis, monitoring, treatment, alleviation of or compensation for an injury or handicap.
- Investigation, replacement, or modification of the anatomy or of a physiological process.
- Control of conception, and which does not achieve its principal intended action in or on the human body by pharmacological, immunological, or metabolic means, but which may be assisted in its function by such means;

b. 'Accessory' means an article which whilst not being a device is intended specifically by its manufacturer to be used together with a device to enable it to be used in accordance with the use of the device intended by the manufacturer of the device;

c. '*In vitro* diagnostic (IVD) medical device' means any medical device which is a reagent, reagent product, calibrator, control material, kit, instrument, apparatus, equipment, or system, whether used alone or in combination, intended by the manufacturer to be used *in vitro* for the examination of specimens, including blood and tissue donations, derived from the human body, solely or principally for the purpose of providing information:

- Concerning a physiological or pathological state;
- Concerning a congenital abnormality;
- To determine the safety and compatibility with potential recipients; or
- To monitor therapeutic measures.

Specimen receptacles are considered to be IVD medical devices. 'Specimen receptacles' are those devices, whether vacuum-type or not, specifically intended by their manufacturers for the primary containment and preservation of specimens derived from the human body for the purpose of IVD examination.

Products for general laboratory use are not IVD medical devices unless such products, in view of their characteristics, are specifically intended by their manufacturer to be used for IVD examination;

d. 'Custom-made device' means any device specifically made in accordance with a duly qualified medical practitioner's written prescription which gives, under his responsibility, specific design characteristics and is intended for the sole use of a particular patient.

The abovementioned prescription may also be made out by any other person authorized by virtue of his professional qualifications to do so.

Mass-produced devices which need to be adapted to meet the specific requirements of the medical practitioner or any other professional user shall not be considered to be custom-made devices;

e. 'device intended for clinical investigation' means any device intended for use by a duly qualified medical practitioner when conducting investigations as referred to in Section 2.1 of Annex X in an adequate human clinical environment.

For the purpose of conducting clinical investigation, any other person who, by virtue of his professional qualifications, is authorized to carry out such investigation shall be accepted as equivalent to a duly qualified medical practitioner;

f. 'manufacturer' means the natural or legal person with responsibility for the design, manufacture, packaging, and labeling of a device before it is placed on the market under his own name, regardless of whether these operations are carried out by that person himself or on his behalf by a third party.

The obligations of this Directive to be met by manufacturers also apply to the natural or legal person who assembles, packages, processes, fully refurbishes and/or labels one or more ready-made products and/or assigns to them their intended purpose as a device with a view to their being placed on the market under his own name. This subparagraph does not apply to the person who, while not a manufacturer within the meaning of the first subparagraph, assembles or adapts devices already on the market to their intended purpose for an individual patient;

g. 'intended purpose' means the use for which the device is intended according to the data supplied by the manufacturer on the labeling, in the instructions and/or in promotional materials;

h. 'placing on the market' means the first making available in return for payment or free of charge of a device other than a device intended for clinical investigation, with a view to distribution and/or use on the Community market, regardless of whether it is new or fully refurbished;

i. 'putting into service' means the stage at which a device has been made available to the final user as being ready for use on the Community market for the first time for its intended purpose;

j. 'authorized representative' means any natural or legal person established in the Community who, explicitly designated by the manufacturer, acts, and maybe addressed by authorities and bodies in the Community instead of the manufacturer with regard to the latter's obligations under this Directive;

k. 'clinical data' means the safety and/or performance information that is generated from the use of a device. Clinical data are sourced from:

- Clinical investigation(s) of the device concerned;
- Clinical investigation(s) or other studies reported in the scientific literature, of a similar device for which equivalence to the device in question can be demonstrated; or
- Published and/or unpublished reports on other clinical experience of either the device in question or a similar device for which equivalence to the device in question can be demonstrated.

l. 'Device subcategory' means a set of devices having common areas of intended use or common technology;

m. 'Generic device group' means a set of devices having the same or similar intended uses or commonality of technology allowing them to be classified in a generic manner not reflecting specific characteristics;

n. 'Single-use device' means a device intended to be used once only for a single patient.

3. Where a device is intended to administer a medicinal product within the meaning of Article 1 of Directive 2001/83/EC (1), that device shall be governed by this Directive, without prejudice to the provisions of Directive 2001/83/EC with regard to the medicinal product.

If, however, such a device is placed on the market in such a way that the device and the medicinal product form a single integral product which is intended exclusively for use in the given combination and which is not reusable, that single product shall be governed by Directive 2001/83/EC. The relevant ER of Annex I to this Directive shall apply as far as safety and performance-related device features are concerned.

4. Where a device incorporates, as an integral part, a substance which, if used separately, may be considered to be a medicinal product within the meaning of Article 1 of Directive 2001/83/EC and which is liable to act upon the body with action ancillary to that of the device, that device shall be assessed and authorized in accordance with this Directive.

4a. Where a device incorporates, as an integral part, a substance which, if used separately, may be considered to be a medicinal product constituent or a medicinal product derived from human blood or human plasma within the meaning of Article 1 of Directive 2001/83/EC and which is liable to act upon the human body with action ancillary to that of the device, hereinafter referred to as a 'human blood derivative,' that device shall be assessed and authorized in accordance with this Directive.

5. This Directive shall not apply to:

a. In vitro diagnostic devices;

b. Active implantable devices covered by Directive 90/385/EEC;

c. Medicinal products covered by Directive 2001/83/EC. In deciding whether a product falls under that Directive or this Directive, particular account shall be taken of the principal mode of action of the product;

d. Cosmetic products covered by Directive 76/768/EEC (1); e) human blood, blood products, plasma, or blood cells of human origin or to devices which incorporate at the time of placing on the market such blood products, plasma, or cells, with the exception of devices referred to in paragraph 4a;
e. Transplants or tissues or cells of human origin nor to products incorporating or derived from tissues or cells of human origin, with the exception of devices referred to in paragraph 4a;
f. Transplants or tissues or cells of animal origin, unless a device is manufactured utilizing animal tissue which is rendered non-viable or non-viable products derived from animal tissue.
6. Where a device is intended by the manufacturer to be used in accordance with both the provisions on personal protective equipment in Council Directive 89/686/EEC (1) and this Directive, the relevant basic health and safety requirements of Directive 89/686/EEC shall also be fulfilled.
7. This Directive is a specific Directive within the meaning of Article 1(4) of Directive 2004/108/EC of the European Parliament and of the Council (2).
8. This Directive shall not affect the application of Council Directive 96/29/Euratom of 13 May 1996 laying down basic safety standards for the protection of the health of workers and the general public against the dangers arising from ionizing radiation (3), nor of Council Directive 97/43/Euratom of 30 June 1997 on health protection of individuals against the dangers of ionizing radiation in relation to medical exposure.

The medical devices directive outlines the minimum requirements a manufacturer should meet when aiming to ensure the safety and performance characteristics for medical devices in the European market. It covers the following principles:

- The ER: this part of the directive outlines the necessary precautions and requirements to be considered in the design, manufacture, use, and disposal of medical devices.
- Classification of devices and conformity assessment procedures which are based on the level of risk inherent in the device.
- Control of production to ensure conformity to type.
- Monitoring and vigilance.

Manufacturers of Medical Devices should demonstrate conformity to all requirements listed within the Directive in addition to the requirements in subsequent guideline documents such as MEDDEVs (MDDs).

For many of the common techniques and product types, there exist what is known as harmonized standards. While use of a harmonized standard is not always mandatory, it is highly recommended as it represents best practice and technical state of the art, which can be used as a presumption of conformity to the relevant part of the directive globally. A manufacturer may however choose to not follow a harmonized standard, which is also okay, but, in this case, they must be able to demonstrate clearly how the product meets the relevant safety or performance requirements.

Article 5 of the MDD 93/42/EEC: Reference to Standards
1. Member States shall presume compliance with the ER referred to in Article 3 in respect of devices which are in conformity with the relevant national standards adopted pursuant to the harmonized standards the references of which have been publishes in the Official Journal of the European Communities; Member States shall publish the references of such national standards. 2. For the purposes of this Directive, reference to harmonized standards also includes the monographs of the European Pharmacopoeia notably on surgical sutures and on interaction between medicinal products and materials used in devices containing such medicinal products, the references of which have been published in the Official Journal of the European Communities. 3. If a Member State or the Commission considers that the harmonized standards do not entirely meet the ER referred to in Article 3, the measures to be taken by the Member States with regard to these standards and the publication referred to in paragraph 1 of this Article shall be adopted by the procedure defined in Article 6(2).

5.2.4. Understanding Which Products Are Not Covered by the Medical Device Directive (MDD)

The directive does not apply to the following:

a. In vitro diagnostic devices;

b. Active implantable devices covered by Directive 90/385/EEC;

c. Medicinal products covered by Directive 65/65/EEC, including medicinal products derived from blood as covered by Directive 89/381/EEC;

d. Cosmetic products covered by Directive 76/768/EEC;

e. Human blood, blood products, plasma, or blood cells of human origin or to devices which incorporate at the time of placing on the market such blood products, plasma, or cells, with the exception of devices referred to in paragraph 4a;

f. Transplants or tissues or cells neither of human origin nor to products incorporating or derived from tissues or cells of human origin;

g. Transplants or tissues or cells of animal origin, unless a device is manufactured utilizing animal tissue which is rendered non-viable or non-viable products derived from animal tissue.

5.2.5. Purpose of the Medical Device Directive (MDD) 93/42/EEC

The Medical Devices Directive was enforced to provide a harmonized regulatory environment for all medical devices sold within the European Union. All products which fall within the scope of the Directive must meet certain essential safety and administrative requirements before being CE marked to show compliance. Such products may then be freely sold within the EU without being subject to any additional national regulations.

5.2.6. Scope of the Medical Device Directive (MDD) 93/42/EEC

The medical devices directive is one of three directives which together cover all medical equipment. The remaining two directives are: The active implantable medical devices Directive (AIMDD) and the *in vitro* diagnostic devices directive (IVDD).

Let us look at how the Directive defines medical devices. According to the MDD: a medical device is defined as:

"Any instrument, apparatus, appliance, material or other article, whether used alone or in combination, including the software necessary for its proper application intended by the manufacturer to be used for human beings for the purpose of:

- Diagnosis, prevention, monitoring, treatment, or alleviation of disease.
- Diagnosis, monitoring, treatment, or alleviation of or compensation for an injury or handicap.
- Investigation, replacement, or modification of the anatomy or of a physiological process.
- Control of conception and which does not achieve its principal intended action in or on the human body by pharmacological, immunological, or metabolic means, but which may be assisted in its function by such means."

The scope of the MDD is quite wide, and in some instances, there may be some form of overlap with other Directives. The MDD does however include provisions for all relevant hazards such as mechanical and electrical safety as well as provisions for electromagnetic compatibility. Devices which fall within the scope of the MDD, the AIMDD, and the IVDD are generally excluded from the scope of other Directives. It is also important for a manufacturer to realize that some devices/equipment may fall into or out of the scope of the MDD solely on the basis of the purpose to which the equipment is intended to be used and the way it is marketed. In simpler terms, if a manufacturer makes any claims that a piece of equipment has some form of therapeutic effect or which alleviates the symptoms of any disability, injury, or illness, then it falls within the scope of the MDD and a product registration is required as well as appropriate documentation. Common examples of products which are considered to be medical devices include:

- Medicine measuring cups;
- Syringes;
- Dental instruments;
- Stethoscopes;
- Thermometers;
- Prescription spectacles and contact lenses;
- Bandages and splints;
- Dental treatment chairs;
- Wheelchairs;
- Condoms;
- First aid kits.

Chapter

6

The Medical Device Regulation MDR 2017/745

CONTENTS

6.1. THE MEDICAL DEVICE REGULATION (EU MDR 2017/745)

6.1.1. What Is the Medical Device Regulation (EU MDR 2017/745 or MDR)?

The medical device regulation (MDR) was published on May 26, 2017, with the aim of replacing the medical device directive (MDD) and the Active Implantable Medical Devices Directive. The MDR will replace the current EU MDD (93/42/EEC) and the EU's Directive on active implantable medical devices (90/385/EEC). All currently certified medical devices and active implantable medical devices (AIMD) must be recertified such that they comply with the new requirements of the MDR. The MDR was meant to come into effect on May 26, 2020, but this has now been delayed for a year to May 2021. The *in-vitro* diagnostic regulation (IVDR) replaced the IVD Directive, and comes into effect on May 26, 2022. The new MDR provides an additional time after the date of application allowing placing new products under the MDD for a grace period.

6.1.2. Major Differences between the Medical Device Directive (MDD) 93/42/EEC and Medical Device Regulation (MDR) 2017/745

The basic structure and requirements of the Regulation remain the same as that of the MDD however a summary of the major differences between the MDD 93/42/EEC and the MDR 2017/745 is provided below:

- There is a wider and more detailed scope of EU legislation which has been extended to include a greater variety of devices such as implants for esthetic purposes.
- There is much stronger supervision of notified bodies (NBs) by national authorities.
- The rights and responsibilities for manufacturers, authorized representatives (AR), importers, and distributors have been defined more clearly.
- There is an extended database on medical devices (known as Eudamed) which aims to provide a more comprehensive and public information source on devices that are available on the EU market.

- More stringent rules for clinical investigations on devices.
- Introduction of classification rules that divide the broad range of *in vitro* diagnostic (IVD) medical devices into four different risk classes as it already exists for other medical devices.
- Creation of a medical device coordination group composed of members representing national competent authorities (CA) in the field of medical devices to ensure better coordination between member states.

Chapter 7

Overview of Medical Device Classification

CONTENTS

7.1. CLASSIFICATION OF MEDICAL DEVICES

Medical devices are generally divided into different classes based on their use and risk when in contact with a patient. This is done in an attempt to ensure the correct regulations and quality systems can be applied with the ultimate goal of ensuring safe and efficient use in or on patients. Some countries have their own separate classification system for different groups of medical devices, such as general devices, active implantable medical devices (AIMD), and *in vitro* diagnostic (IVD) medical devices, while other countries utilize one system for classification of all medical devices. Regardless of the country or system used, all classification schemes tend to be risk based. Having a suitable classification system in place ensures a more appropriate and graduated system of control which allows manufacturers themselves to classify their own devices rather than by one or two general committees, which in effect would take too long. This also enforces those manufacturers should be able to know as early as possible in which class their product falls. In the United States of America, for example, medical devices are classified as Class I, II, and III, with Class III devices representing the highest risk type device and requiring more overall control. Medical devices regulated with the USA are classified through an FDA classification database and are given a seven-digit number based on the category of the product. On the other hand, in the European Union, general devices are classified as Class I, Class I sterile, Class I measuring, Class IIa, Class IIb, or Class III; with Class III devices representing the highest risk device. In addition, IVD devices have their own classification system while active implantable devices are not classified at all.

While having such a classification system in place is essential, it should also be noted that due to the vast number of variables involved as well as the changing nature of some of these variables, the mere implementation of a simple set of classification rules solely based on technical features is impossible. As a result, specifically in the EU, the European legislator has since established a classification concept which is based on the potential hazard related to the use and possible failure of various medical devices while at the same time taking into account technology used and any related standards, regulations, and/or health policy considerations. This has resulted in the establishment of a small set of criteria; namely the duration of contact with the body, degree of invasiveness, and local vs systemic effects, which can be combined in various ways to determine the class of the product. In addition, despite these various classification systems or existing rules which aim to classify the vast majority of medical devices, there are however a small number of exceptions or confusing classification situations which may arise. Common examples include borderline devices between two classes, devices that cannot be classified by existing rules due to their unusual nature, or even situations in which classification may lead to the wrong conformity assessment level being chosen.

Due to the varying classification systems, the GHTF has developed a recommended classification system in which medical devices are divided into classes, namely A, B, C, and D, with Class D representing the highest risk class. This global scheme is however a recommendation to regulatory authorities (Table 7.1).

Table 7.1: Proposed General Classification System for Medical Devices

Class	Risk Level	Device Examples
A	Low risk	Surgical retractors/tongue depressors
B	Low-moderate risk	Hypodermic needles/suction equipment
C	Moderate-high risk	Lung ventilator/bone fixation plate
D	High risk	Heart valves/implantable defibrillator

Source: GHTF (2006).

Figure 7.1 shows an overall illustration of increasing levels of regulatory requirements as the device risk class increases. These regulatory controls may include (1) operation of a quality system which is recommended for all medical devices; (2) technical data; (3) product testing using in-house or independent resources; (4) documentation of clinical evidence to support the manufacturer's claims; (5) the need for and frequency of independent external audit of the manufacturer's quality system; and (6) independent external review of the manufacturer's technical data.

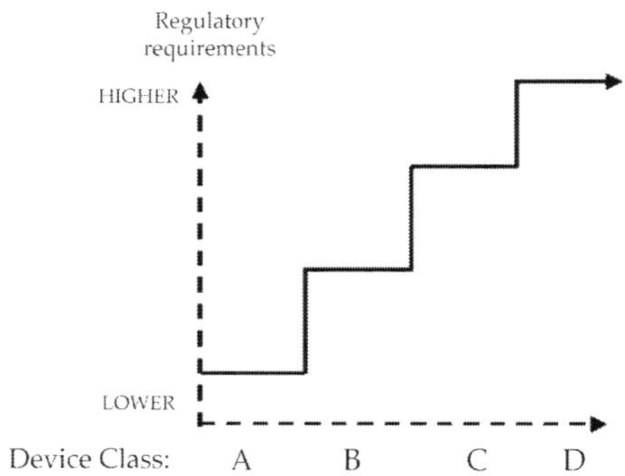

Figure 7.1: Illustration of various regulatory controls as per increasing risk class of medical devices.

7.1.1. Factors Influencing Device Classification Schemes

Several factors are known to influence medical device classification schemes. These range from the duration of contact which the device has with the body to the overall degree of invasiveness to whether the medical device is intended to have a biological or systemic effect on the patient. In some instances, the way in which a manufacturer defines the devices intended use may result in two or more classification rules being applicable to the device itself. In this instance, the highest level of classification will apply. If one medical device is intended to be used together with another medical device that may or may not be from the same manufacturer, then the classification rules should apply separately to each of the devices involved; for example, a general-purpose suture and a suture holder. In some instances, medical devices may be grouped and packaged together as per a manufacturer's requirements, and in this instance, classification of the medical device assemblage will depend on the manufacturer's purpose in packaging and marketing of such medical device combinations as when compared to individually packaged devices. For example, if the combination of devices together results in an overall product that is intended by the manufacturer to meet a specific purpose that is different from that of the individual medical devices that make up the combination, then the combination itself becomes a new medical device in its own right and should be classified according to the new intended use. However, on the other hand, if the medical device combination is for the convenience of the user but does not change the intended uses of the individual medical devices that make up that combination, then the classification allocated to that assemblage is at the level of the highest classified device within that group. On the other hand, if one or more medical devices within that assemblage of devices has yet to comply with all relevant regulatory requirements, then the combination should be classified based on the intended use of the device.

It is also important to remember that medical accessories that have been intended specifically to be used together with a 'parent' medical device such that it enables that medical device to achieve its intended purpose, should be subjected to all the GHTF guidance documents that apply to medical devices to ensure proper classification and application of relevant regulatory rules. These include, for example, GHTF documents around Essential principles for Safety and Performance, post-market surveillance, medical device classification, and so on. in many instances, these medical device accessories may be classified as though they form actual medical devices in their own right.

In other instances, while most software required for certain medical device operation may be incorporated into the medical device itself, some are not. If these standalone software falls within the scope of what is defined to be a medical device, then this software should be classified as follows:

- If the software drives or influences the use of a separate medical device, it should be classified according to the intended use of the combination;
- If the software is independent of any other medical device, it should be classified in its own right using the rules defined below (taken from Section 8 of GHTF/SG1/N15:2006 Principles of Medical Device Classification, Study Group 1 as endorsed by The Global Harmonized Task Force, 2006);
- If the standalone software can be defined as being an active device.

7.1.2. Initial Classification Rules as Defined by the Global Harmonization Task Force (GHTF)

The actual classification of each device depends on the intended use claims made by the manufacturer, and this indicates a need for manufacturers to thoroughly understand the purpose of each rule as defined below. For this reason, the actual classification of particular devices should be considered individually, taking into account its design and intended use (Table 7.2).

Table 7.2: Classification Rules as Defined by the GHTF

Rule	Illustrative Examples of Devices That may Conform with a Rule
Non-Invasive Devices	
• **Rule 1:** All non-invasive devices which come into contact with injured skin: • are in Class A if they are intended to be used as a mechanical barrier, for compression or for absorption of exudates only, i.e., they heal by primary intent; • are in Class B if they are intended to be used principally with wounds which have breached the dermis, including devices principally intended to manage the microenvironment of a wound unless they are intended to be used principally with wounds which have breached the dermis and can only heal by secondary intent, in which case they are in Class C.	Devices covered by this rule are extremely claim sensitive. Simple wound dressings; cotton wool, non-medicated impregnated gauze dressings. Devices used to treat wounds where the subcutaneous tissue is as least partially exposed, and the edges of the wound are not sufficiently close to be pulled together. To close the wound, new tissue must be formed within the wound prior to external closure. The device manufacturer claims that they promote healing through physical methods other than 'primary intent.' Examples include dressings for chronic ulcerated wounds; dressings for severe burns.
• **Rule 2:** All non-invasive devices intended for channeling or storing: • body liquids or tissues • liquids or • gases for the purpose of eventual infusion, administration or introduction into the body are in Class A Unless they may be connected to an active medical device in Class B or a higher class, in which case they are Class B; unless they are intended for use of: • channeling blood, or • storing or channeling other body liquids, or • for storing organs, parts of organs or body tissues, in which case they are Class B. Unless they are blood bags, in which case they are Class C.	Such devices are 'indirectly invasive' in that they channel or store liquids that will eventually be delivered into the body (see comment for Rule 4). *Examples*: administration sets for gravity infusion; syringes without needles. *Examples*: syringes and administration set for infusion pumps; anesthesia breathing circuits. *Note*: "Connection" to an active device covers those circumstances where the safety and performance of the active device is influenced by the non-active device and vice versa. *Examples*: tubes used for blood transfusion, organ storage containers *Example*: Blood bags that do not incorporate an anti-coagulant. *Note*: in some jurisdictions, blood bags have a special rule that places them within a different risk class.

• **Rule 3:** All non-invasive devices intended for modifying the biological or chemical composition of: • blood; • other body-liquids; or • other liquids intended for infusion into the body are in Class C. Unless the treatment consists of filtration, centrifuging or exchanges of gas or of heat, in which case they are in Class B.	Such devices are indirectly invasive in that they treat or modify substances that will eventually be delivered into the body (see note to comment for Rule 4). They are normally used in conjunction with an active device within the scope of either Rule 9 or 11. *Examples*: hemodialyzers; devices to remove white blood cells from whole blood. *Note*: for the purpose of this part of the rule, 'modification' does not include simple, mechanical filtration or centrifuging, which are covered below. *Examples*: devices to remove carbon dioxide; particulate filters in an extracorporeal circulation system.
• **Rule 4:** All other non-invasive devices are in Class A.	These devices either do not touch the patient or contact intact skin only. *Examples*: urine collection bottles; compression hosiery; non-invasive electrodes, hospital beds
Invasive Devices	
• **Rule 5:** All invasive devices with respect to body orifices (other than those which are surgically invasive) and which: • are not intended for connection to an active medical device, or • are intended for connection to a Class A medical device only. • are in Class A if they are intended for transient use; • are in Class B if they are intended for short-term use; Unless they are intended for short-term use in the oral cavity as far as the pharynx, in an ear canal up to the eardrum or in a nasal cavity, in which case they are in Class A: • are in Class C if they are intended for long-term use; Unless they are intended for long-term use in the oral cavity as far as the pharynx, in an ear canal up to the eardrum or in a nasal cavity and are not liable to be absorbed by the mucous membrane, in which case they are in Class B. All invasive devices with respect to body orifices (other than those which are surgically invasive) that are intended to be connected to an active medical device in Class B or a higher class, are in Class B.	(Section 4). Devices tend to be diagnostic and therapeutic instruments used in ENT, ophthalmology, dentistry, proctology, urology, and gynecology. Classification depends on the duration of use and the sensitivity (or vulnerability) of the orifice to such invasion. *Examples*: examination gloves; enema devices. *Examples*: urinary catheters, tracheal tubes *Examples*: dentures intended to be removed by the patient; dressings for nose bleeds *Example*: urethral stent; contact lenses for long-term continuous use (for this device, removal of the lens for cleaning or maintenance is considered as part of the continuous use). *Examples*: orthodontic wire, fixed dental prosthesis. *Examples*: tracheal tubes connected to a ventilator; suction catheters for stomach drainage; dental aspirator tips. *Note*: independent of the time for which they are invasive.

□ **Rule 6:** All surgically invasive devices intended for transient use are in Class B • Unless they are reusable surgical instruments, in which case they are in Class A; or • Unless intended to supply energy in the form of ionizing radiation, in which case they are in Class C; or • Unless intended to have a biological effect or be wholly or mainly absorbed, in which case they are in Class C; or • unless intended to administer medicinal products by means of a delivery system, if this is done in a manner that is potentially hazardous taking account of the mode of application, in which they are in Class C; or • unless they are intended specifically for use in direct contact with the central nervous system, in which case they are in Class D; or • unless intended specifically to diagnose, monitor, or correct a defect of the heart or of the central circulatory system through direct contact with these parts of the body, in which case they are in Class D.	A majority of such devices fall into several major groups: those that create a conduit through the skin (e.g., syringe needles; lancets), surgical instruments (e.g., single-use scalpels; surgical staplers; single-use aortic punch); surgical gloves; and various types of catheter/suckers, etc. *Note*: a surgical instrument (other than those in Class D) is in Class A if reusable and in Class B if supplied sterile and intended for single use. Also, a surgical instrument connected to an active device is in a higher class than A. *Note*: if the device incorporates a medicinal substance in a secondary role, refer to Rule 13. *Examples*: Manually operated surgical drill bits and saws. *Example*: catheter incorporating/containing sealed radioisotopes *Notes*: (a) the 'biological effect' referred to is an intended one rather than unintentional. The term 'absorption' refers to the degradation of a material within the body and the metabolic elimination of the resulting degradation products from the body. (b) This part of the rule does not apply to those substances that are excreted without modification from the body. *Example*: Insufflation gases for the abdominal cavity. *Example*: insulin pen for self-administration. *Note*: the term 'administration of medicines' implies storage and/or influencing the rate/volume of medicine delivered not just channeling. The term 'potentially hazardous manner' refers to the characteristics of the device and not the competence of the user. *Examples*: angioplasty balloon catheters and related guide wires; dedicated disposable cardiovascular surgical instruments.
• **Rule 7:** All surgically invasive devices intended for short-term use are in Class • unless they are intended to administer medicinal products, in which case they are in Class C; or • unless they are intended to undergo chemical change in the body (except if the devices are placed in the teeth), in which case they are in Class C; or	Such devices are mostly used in the context of surgery or post-operative care, or are infusion devices, or are catheters of various types. *Examples*: infusion cannula; temporary filling materials; non-absorbable skin closure devices; tissue stabilizers used in cardiac surgery. *Note*: includes devices that are used during cardiac surgery but do not monitor or correct a defect. *Note*: if the device incorporates a medicinal substance in a secondary role, refer to Rule 13.

• unless they are intended to supply energy in the form of ionizing radiation, in which case they are in Class C; or • unless they are intended to have a biological effect or to be wholly or mainly absorbed, in which case they are in Class D; or • unless they are intended specifically for use in direct contact with the central nervous system, in which case they are in Class D; • unless they are intended specifically to diagnose, monitor, or correct a defect of the heart or of the central circulatory system through direct contact with these parts of the body, in which case they are in Class D.	*Note*: the term 'administration of medicines' implies storage and/or influencing the rate/volume of medicine delivered not just channeling. *Example*: surgical adhesive. *Example*: brachytherapy device *Example*: absorbable suture; biological adhesive. *Note*: the 'biological effect' referred to is an intended one rather than unintentional. The term 'absorption' refers to the degradation of a material within the body and the metabolic elimination of the resulting degradation products from the body. *Example*: neurological catheter. *Examples*: cardiovascular catheters; temporary pacemaker leads; carotid artery shunts.
• **Rule 8:** All implantable devices, and long-term surgically invasive devices, are in Class C: • unless they are intended to be placed into the teeth, in which case they are in Class B; or • unless they are intended to be used in direct contact with the heart, the central circulatory system, or the central nervous system, in which case they are in Class D; or • unless they are intended to be life-supporting or life-sustaining, in which case they are in Class D; or • unless they are intended to be active implantable medical devices, in which case they are Class D; or • unless they are intended to have a biological effect or to be wholly or mainly absorbed, in which case they are in Class D; or • unless they are intended to administer medicinal products, in which case they are in Class D; or • unless they are intended to undergo chemical change in the body (except if the devices are placed in the teeth), in which case they are in Class D; or • unless they are breast implants, in which case they are in Class D.	Most of the devices covered by this rule are implants used in the orthopedic, dental, ophthalmic, and cardiovascular fields. *Example*: maxilla-facial implants; prosthetic joint replacements; bone cement; non-absorbable internal sutures; posts to secure teeth to the mandibular bone (without a bioactive coating). *Note*: if the device incorporates a medicinal substance in a secondary role, refer to Rule 13. *Examples*: bridges; crowns; dental filling materials. *Examples*: prosthetic heart valves; spinal and vascular stents. *Example*: pacemakers, their electrodes, and their leads; implantable defibrillators. *Example*: implants claimed to be bioactive. *Note*: hydroxy-apatite is considered as having biological effect only if so claimed and demonstrated by the manufacturer. *Example*: rechargeable non-active drug delivery system. *Note*: bone cement is not within the scope of the term 'chemical change in the body' since any change takes place in the short rather than long term.

Active Devices	
• **Rule 9(i):** All active therapeutic devices intended to administer, or exchange energy are in Class B • unless their characteristics are such that they may administer or exchange energy to or from the human body in a potentially hazardous way, including ionizing radiation, taking account of the nature, the density and site of application of the energy, in which case they are in Class C.	Such devices are mostly electrically powered equipment used in surgery; devices for specialized treatment and some stimulators. *Examples*: muscle stimulators; TENS devices; powered dental handpieces; hearing aids; neonatal phototherapy equipment; ultrasound equipment for physiotherapy. *Examples*: lung ventilators; baby incubators; electrosurgical generators; external pacemakers and defibrillators; surgical lasers; lithotripters; therapeutic X-ray and other sources of ionizing radiation. *Note*: the term 'potentially hazardous' refers to the type of technology involved and the intended application.
• **Rule 9(ii):** All active devices intended to control or monitor the performance of active therapeutic devices in Class C, or intended directly to influence the performance of such devices, are in Class C.	*Examples*: external feedback systems for active therapeutic devices.
• **Rule 10(i):** Active devices intended for diagnosis are in Class B: • if they are intended to supply energy which will be absorbed by the human body (except for devices used solely to illuminate the patient's body, with light in the visible or near infra-red spectrum, in which case they are Class A), or • if they are intended to image *in vivo* distribution of radiopharmaceuticals, or • if they are intended to allow direct diagnosis or monitoring of vital physiological processes Unless they are specifically intended for: • monitoring of vital physiological parameters, where the nature of variations is such that it could result in immediate danger to the patient, for instance variations in cardiac performance, respiration, activity of central nervous system, or • diagnosing in clinical situations where the patient is in immediate danger, in which case they are in Class C	Such devices include equipment for ultrasonic diagnosis/imaging, capture of physiological signals, interventional radiology, and diagnostic radiology. *Examples*: magnetic resonance equipment; diagnostic ultrasound in non-critical applications; evoked response stimulators. *Examples*: gamma/nuclear cameras *Example*: electronic thermometers, stethoscopes, and blood pressure monitors; electrocardiographs. *Example*: monitors/alarms for intensive care; biological sensors; oxygen saturation monitors; Apnea monitors. *Example*: ultrasound equipment for use in interventional cardiac procedures.

•••• **Rule 10(ii):** Active devices intended to emit ionizing radiation and intended for diagnostic and/or interventional radiology, including devices which control or monitor such devices, or those which directly influence their performance, are in Class C.	*Example*: these include devices for the control, monitoring or influencing of the emission of ionizing radiation.
• **Rule 11:** All active devices intended to administer and/or remove medicinal products, body-liquids or other substances to or from the body are in Class B Unless this is done in a manner that is potentially hazardous, taking account of the nature of the substances involved, of the part of the body concerned and of the mode and route of administration, in which case they are in Class C.	Such devices are mostly drug delivery systems or anesthesia equipment. Examples of Class B devices: suction equipment; feeding pumps; jet injectors for vaccination; nebulizer to be used on conscious and spontaneously breathing patients where failure to deliver the appropriate dosage characteristics is not potentially hazardous. *Examples*: infusion pumps; anesthesia equipment; dialysis equipment; hyperbaric chambers; nebulizer where the failure to deliver the appropriate dosage characteristics could be hazardous.
• **Rule 12:** All other active devices are in Class A	*Examples*: examination lamps; surgical microscopes; powered hospital beds and wheelchairs; powered equipment for the recording, processing, viewing of diagnostic images; dental curing lights.
Additional Rules	
• **Rule 13:** All devices incorporating, as an integral part, a substance which, if used separately, can be considered to be a medicinal product, and which is liable to act on the human body with action ancillary to that of the devices, are in Class D.	These medical devices incorporate medicinal substances in an ancillary role. *Examples*: antibiotic bone cements; heparin-coated catheters; wound dressings incorporating antimicrobial agents to provide ancillary action on the wound; blood bags incorporating an anti-coagulant. *Note*: Such medical devices may be subject to additional conformity assessment procedures according to the regional or national requirements of medicinal product Regulatory Authorities.
• **Rule 14:** All devices manufactured from or incorporating animal or human cells/tissues/derivatives thereof, whether viable or non-viable, are Class D Unless such devices are manufactured from or incorporate non-viable animal tissues or their derivatives that come in contact with intact skin only, where they are in Class A.	*Note*: In some jurisdictions such products: -are considered to be outside the scope of the medical device definition; • may be subject to different controls It is likely the regulations controlling these devices will be the subject of future harmonization efforts. *Examples*: porcine heart valves; catgut sutures *Examples*: leather components of orthopedic appliances.

• **Rule 15:** All devices intended specifically to be used for sterilizing medical devices, or disinfecting as the endpoint of processing, are in Class C. • unless they are intended for disinfecting medical devices prior to endpoint sterilization or higher-level disinfection, in which case they are in Class B; or • unless they are intended specifically to be used for disinfecting, cleaning, rinsing or, when appropriate, hydrating contact lenses, in which case they are in Class C.	*Examples*: devices for disinfecting or sterilizing endoscopes; disinfectants intended to be used with medical devices. *Note*: This rule does not apply to products that are intended to clean medical devices by means of physical action, e.g., washing machines. *Example*: washer-disinfectors In some jurisdiction's solutions for use with contact lenses: -are considered to be outside the scope of the medical device's definition; -may be subject to different controls.
• **Rule 16:** All devices used for contraception or the prevention of the transmission of sexually transmitted diseases are in Class C Unless they are implantable or long-term invasive devices, in which case they are in Class D.	*Examples*: condoms, contraceptive diaphragms *Example*: intrauterine contraceptive device

Chapter

8

Understanding the EU Guidelines on Medical Device Classification

CONTENTS

8.1. EU GUIDELINES ON MEDICAL DEVICE CLASSIFICATION

Detailed explanations of the European classification systems can be found in MEDDEV 2.4/1. General requirements of this classification scheme are such that all devices must:

- Meet the essential requirements (ER) irrespective of the class of the device;
- Be subject to the reporting requirements under the medical device vigilance system;
- Be CE marked (except for custom-made devices and devices that are intended for clinical investigation).

This medical device classification systems are needed to channel medical devices into the proper conformity assessment route (Table 8.1) as subjecting all existing medical devices to rigorous conformity assessment procedures would just not be feasible economically.

Table 8.1: Conformity Assessment Procedures vs. Classes

Conformity Assessment Procedures	**Classes**					
Annexes	**I**	**I Sterile**	**I Measuring**	**IIA**	**IIB**	**III**
II (+ sect. 4)	–	–	–	–	–	√
II (– sect. 4)	–	–	–	√	√	–
III	–	–	–	–	√	√
IV	–	√	√	√	√	√
V	–	√	√	√	√	√
VI	–	√	√	√	√	–
VII	√	√	√	√	–	–

When carrying out a classification as per MEDDEV 2.4/1, the manufacturer should first decide if the product is a medical device as defined by Directive 93/42 or if it is an accessory to a medical device. Determining the exact grouping of the device will aid a manufacturer in knowing which directive to apply. Active implantable devices and devices for *in vitro* diagnosis are covered by separate directives which in actual fact do not apply the classification rules from guidelines supplied by MEDDEV 2.4/1, thereby further indicating that different groups of devices will utilize different rules.

Classification rules as defined by MEDDEV 2.4/1 are based on terms related to: (i) the duration of contact with the patient; (ii) degree of invasiveness; and (iii) the part of the body affected by the use of the device itself. One of the first steps to applying the classification rules is understanding the basic definitions used in the guideline, all of which are defined in Section I of Annex IX of the Directive.

8.1.1. Duration

The duration of contact with the patient, also known as *time* can be classified into transient, short term or long term. Transient is defined as that which is normally intended for continuous use for *less than 60 minutes* while short term is defined as that which is normally intended for continuous use for *not more than 30 days* and long term is that which is normally intended for continuous use *for more than 30 days*.

8.1.2. Continuous Use

In addition, understanding the concept of continuous use is important when the intended use of the device is taken into consideration. The abovementioned concept of duration (transient, short term and long term) is all defined in terms of continuous use, which is better defined as an uninterrupted actual use for the intended purpose of the device in question. For example, a scalpel which may be used on a patient for cutting of tissue will not normally last for more than a few seconds at a time and can therefore be defined as a continuous transient use device. On the other hand, a device that is used and replaced immediately either by the same or identical device is considered as discontinued and shall be defined as an extension of the continuous use of the device. A common example would be a suture needle which is used for stitching wounds.

8.1.3. Invasiveness

Invasive devices are defined as a device which either, completely, or partly, penetrates inside the body. This usually occurs either through a body orifice (better known as any natural opening in the body) or through the surface of the body. An invasive device which penetrates inside the body through the surface of the body, with the aid or in the context of a surgical operation is better known as a surgically invasive device. Thus, a surgically invasive device always implies that it enters through an artificially created opening, which could be a large opening, such as a surgical incision, or as small as a needle hole created by a suture needle.

8.1.4. Implantable Device

An implantable device is defined as any device which is intended for complete introduction into the human body or which will be replaced by an epithelial surface or the surface of the eye; either by surgical intervention and which is intended to remain in place after the procedure. In addition, any device intended to be partially introduced into the human body through surgical intervention and intended to remain in place after the procedure for at least 30 days is also considered an implantable device. When defining an implantable device, the idea behind the term 'procedure' must be taken into consideration and as such, an implantable device must remain in the patient after the procedure is complete. As such, a procedure must include the surgical activity during which the implantable device is placed in the body and the immediate post-operative procedures associate with the initial procedure itself. That being said, the term procedure does not include the removal of an implant, which in effect must be considered to be another procedure. As a result, the placing of an implant and the removal of the implant would be become two different surgical procedures. Furthermore, some partially implanted devices are also classified to be implants, but clear distinction is needed.

8.1.5. Active Devices

Active medical devices are defined as any medical device in which their operation depends on a source of electrical energy or any power source other than that directly generated by the human body or gravity and which acts by converting this energy. Those medical devices that aim to simply transmit energy, substances or other elements between an active medical device and the patient themselves, without any significant change, are not considered to be active medical devices. In order for this definition to be conveyed properly, the concept of "act by converting energy" and "significant changes" must be clearly explained and understood. To "act by converting energy" includes the conversion of energy in the device and/or conversion at the interface between the device itself and the tissues or within the tissues. The concept of "significant changes" on the other hand, includes changes in the nature, level, and density of energy. For example, an electrode would not be defined as an active medical device as per this definition as long as the energy input is intended to be the same as the energy output. On the other hand, electrodes used in electrosurgery for cutting tissues can be defined as active medical devices as per the above definition as their operation depends on energy provided by a generator and their action is achieved by conversion of the energy at the interface between the device and the tissue itself or within the tissue. However, those electrodes that are intended for E.C.G or E.E.G's are to normally define as active devices because they do not normally act by conversion of energy. Manufacturers should however consult the "Guidelines relating to the application of the Council Directive 90/385/EEC on active implantable medical devices (AIMD) (Med. Dev. 2.1/2)" should they require more information or better understanding on the issue.

8.1.6. Devices with a Measuring Function

Devices are considered to have a measuring function if they measure a parameter such as volume or a physiological parameter and display or indicate its value in a defined unit of measurement. Common examples include urine bags or thermometers.

Certain criteria have been laid out as per MEDDEV 2.1/5 (1998) which a Guidance document relating to the application of medical devices with a measuring function. This guidance document states criteria which if fulfilled together, indicate that a device does, in fact, have a measuring function. These criteria are defined below:

- The device is intended by the manufacturer to measure:
  - Quantitatively a physiological or anatomical parameter;
  - A quantity or a quantifiable characteristic of energy or of substances delivered to or removed from the human body.
- The result of the overall measurement:
  - is displayed in legal units or other acceptable units within the meaning of Directive 80/181/ECC;
  - is compared to at least one point of reference indicated in legal units or other acceptable units in compliance with the pre-mentioned directive.
- The intended purpose implies accuracy (claimed either explicitly or implicitly) where any form of non-compliance with the implied accuracy could result in a significant adverse effect on the patient's health and safety.
  - **Note 1:** The phrase "claimed implicitly" covers cases where the user, on the basis of the designation of the device or of its accompanying documents, or on the basis of the common use is entitled to expect accuracy where the accuracy of the measurement has an impact on the diagnosis or therapy of the patient (Table 8.2).

Table 8.2: Common Examples of Devices with a Measuring Function

Criteria	Example
Devices used for measuring body temperature	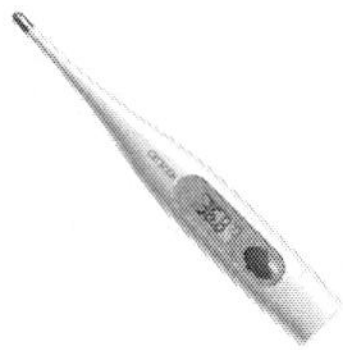
Pacifiers which include a temperature display including those with only a change of color where criteria b is met	
Devices for indicating that body temperature is above or below a specified value	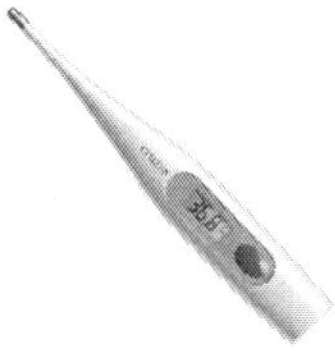
Non-active non-invasive devices for measuring blood pressure	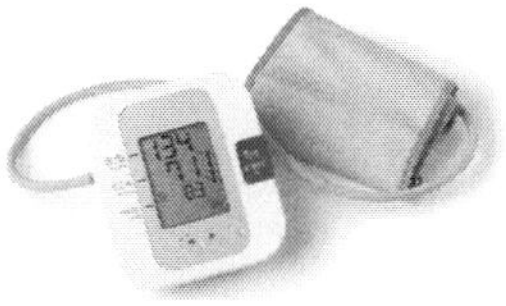

Non-active devices for measuring intra-ocular pressure	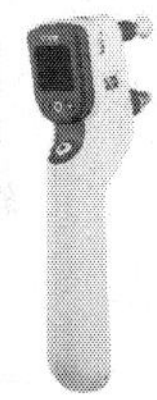
Devices for measuring volume or pressure or flow of liquid or gases delivered to or removed from the human body. This includes any container with a graduation scale or with a single point graduation where criteria c is met.	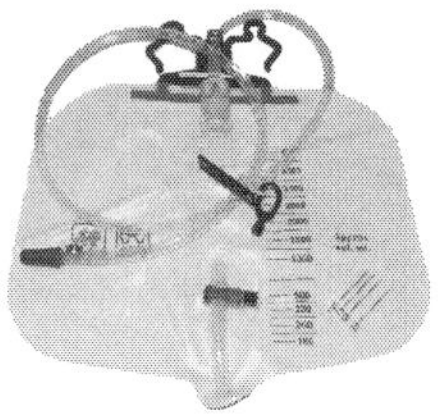

8.1.7. Devices without a Measuring Function

Common examples of devices without a measuring function:

Criteria	Example
Devices used for the delivery of liquid to the human body (for example, medicine spoons, cups, droppers, without graduation or scale or display of measuring unit)	
Devices used for displaying trends of physiological parameters (for example, urine bags without graduation or scale, calipers for obesity)	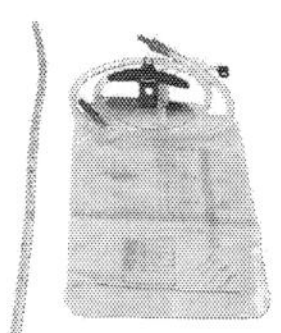
Eye-test charts	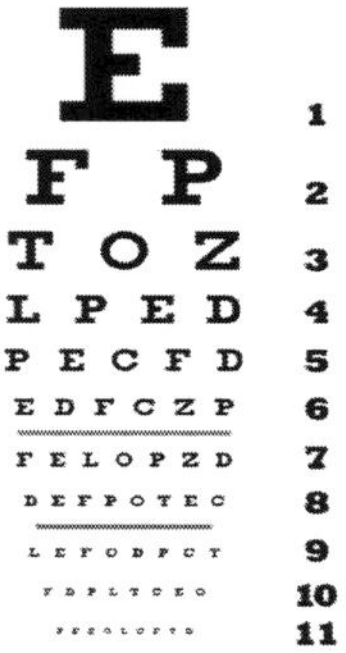

8.2. UNDERSTANDING HOW TO APPLY THE RULES OF MEDICAL DEVICE CLASSIFICATION?

When applying the classification rules, the following should be considered:

- The intended purpose of the device as opposed to technical characteristics as this will determine the class of the device itself (unless these have a direct influence on the intended purpose).
- The intended use must not be confused with common accidental uses of the device.
- The intended purpose assigned by the actual manufacturer to the specific device should determine the class of the device rather than allowing for classification by inferences such as the class of other similar products determining the class of a specific device.
- Classification of device components in relation to the system as a whole or individually should be taken into consideration. For example, combination devices with parts that have different functional purposes may be analyzed separately with respect to each of the parts; such as a drainage device that has an invasive tube and a non-invasive collection device. Each of these components may be classified separately.
- Accessories must be classified separately from their parent devices.
- If a device can be classified by several rules, then the highest possible class should apply.
- If a device is not intended to be used solely in a specific part of the body, then it must be considered and classified based on the most critical specified use. As such classification of the device will have to be determined based on claims contained in the information supplied with the medical device itself. For this reason, manufacturers must be extremely specific with regards to the overall intended use while also providing as a minimum requirement, either appropriate positive or negative indications for use.
- Multi-application equipment (such as laser printers or identification cameras) which may be used in combination with medical devices are not actual medical devices unless their manufacturer places them on the market with specific intended purpose as a medical device or for use with medical devices only.

- Software that is regarded as driving or influencing the use of a medical device may fall in the same class of the device which it influences. If, however, such standalone software is not regarded as driving or influencing the use of a medical device, then it should be classified separately and in its own right.

8.3. HOW TO APPLY THE EU MEDICAL DEVICE CLASSIFICATION RULES?

In order to appropriately apply the EU medical device rules for classification, manufacturers must take into account all the rules laid out in the guidelines to ensure their devices are properly classified. In addition, all device characteristics or combination of characteristics in relation to its intended purpose must be taken into consideration. One such classification example is defined below:

A wound drainage device produced by Manufacturer X needs to be classified—a simple wound drainage device has three components that must be taken into consideration, namely the cannula, the tubing, and the collector unit. If the particular device in question is to be sold without a cannula, then the classification of the cannula does not need to be taken into consideration. The manufacturer has also stated that the device will be used under short term duration, i.e., uninterrupted intended use will be for more than 60 minutes and less than 30 days. It also assumed that any collected liquids are not intended to be re-infused into the body nor reprocessed for eventual re-infusion and that the device is not intended to be connected to a powered suction system (Table 8.3).

Table 8.3: Intended Uses of Devices and Their Risk Classification

Intended Uses	Rule	Class
Surgically invasive cannula to reach a wound site in the pleural cavity to drain the cavity	7	IIA
Non-invasive tubing to evacuate body-liquids towards the collector	1	I
Non-invasive collector to receive the body liquids	1	I

From the above defined specifications by the manufacturer, it is clear that they would have a choice of applying a Class II A risk level to the whole device. Alternatively, they would have to carry out a separate conformity assessment procedure for the cannula and another conformity assessment procedure for the tubing and another for the collector.

As defined by MEDDEV 2.4/1, different rules will apply to different groups of devices (Table 8.4). As a rule of thumb, however, it should be remembered that the characteristics or combination of characteristics in accordance with its intended purpose of the device that determines the highest class will determine the overall class for the device (Figures 8.1–8.5).

Table 8.4: Different Rules That Apply for Risk Classification of Different Devices

Type of Device	Applicable Rules as per MEDDEV 2.4/1 Classification Rules
Non-invasive	1,2,3,4
Invasive	5,6,7,8
Active	9,10,11,12
Special	13,14,15,16,17,18

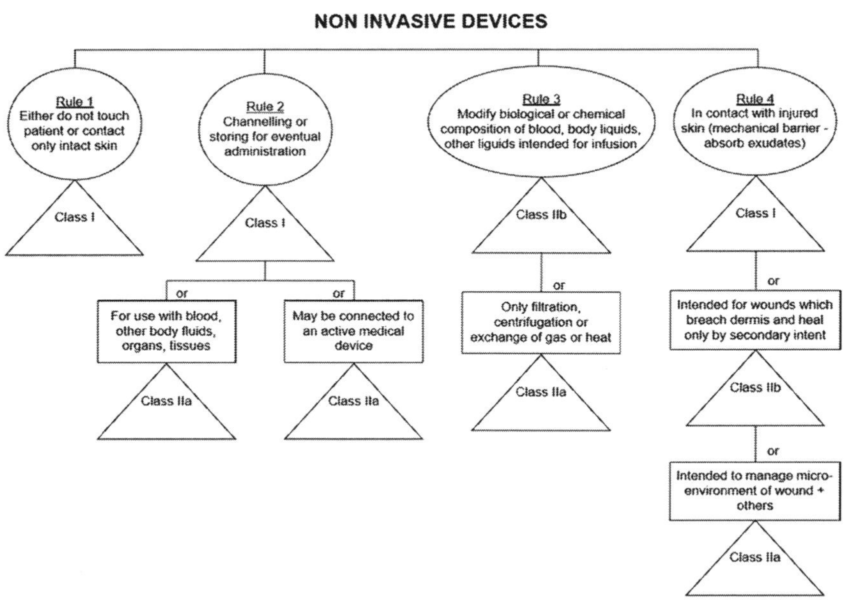

Figure 8.1: Flow diagram showing risk classification routes based on Rules 1–4 from MEDDEV2.4/1.

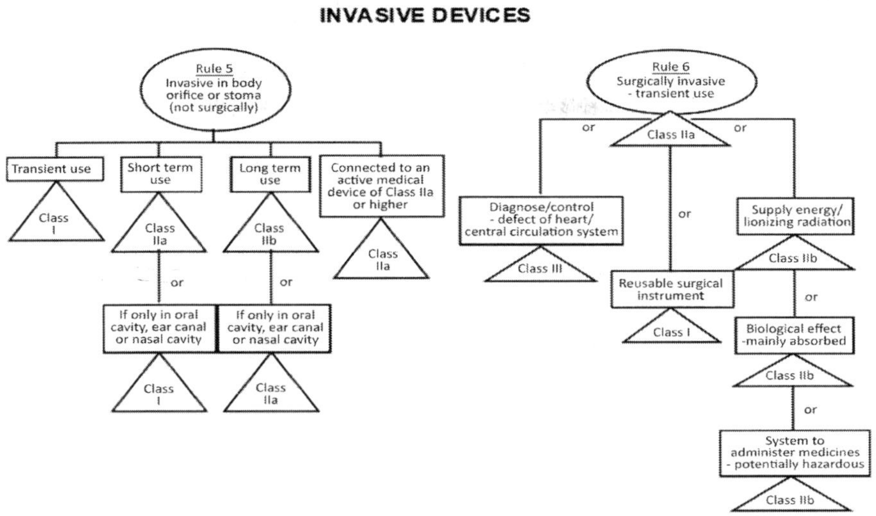

Figure 8.2: Flow diagram showing risk classification routes based on Rules 5–6 from MEDDEV2.4/1.

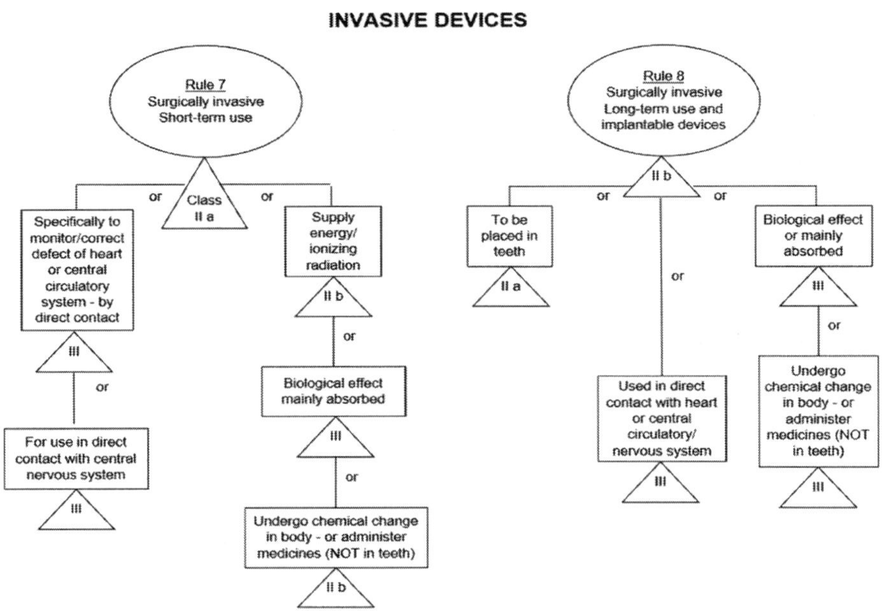

Figure 8.3: Flow diagram showing risk classification routes based on Rules 7–8 from MEDDEV2.4/1.

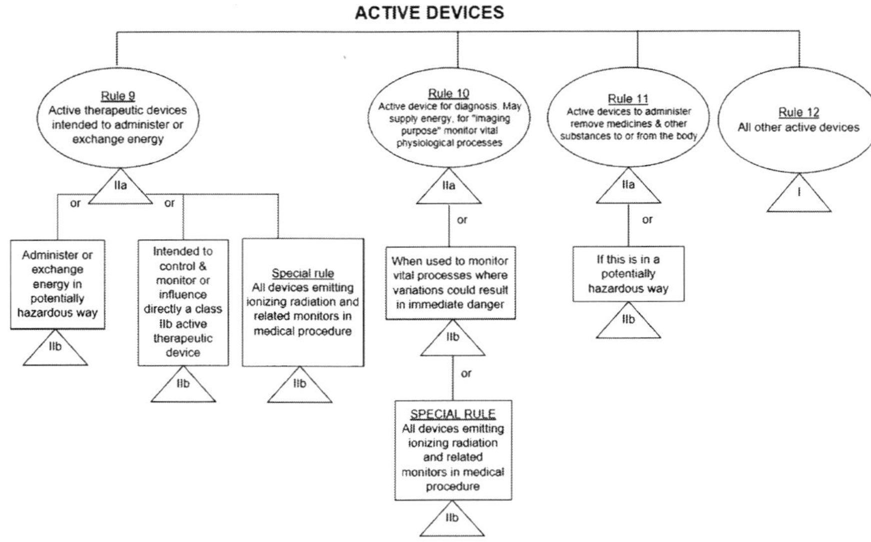

Figure 8.4: Flow diagram showing risk classification routes based on Rules 9–12 from MEDDEV2.4/1.

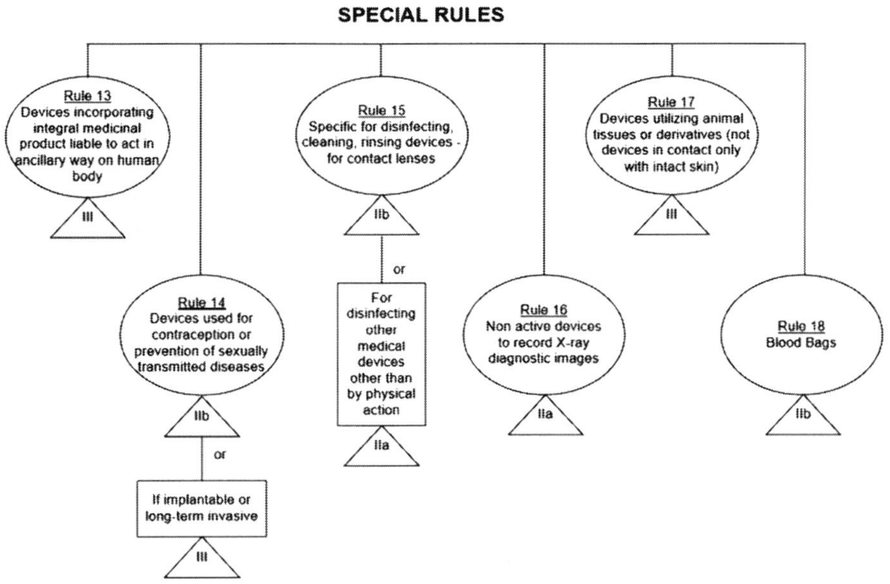

Figure 8.5: Flow diagram showing risk classification routes based on Rules 13–17 from MEDDEV2.4/1.

8.4. UNDERSTANDING THE RULES GOVERNING GUIDELINES FOR THE CLASSIFICATION OF MEDICAL DEVICES ACCORDING TO MEDDEV 2.4/1

The below sub-chapter covers a general explanation of the rules and practical issues governing the EU guidelines for medical device classification as laid out in MEDDEV 2.4/1.

8.4.1. Rule 1: Devices That Either Do Not Touch the Patient or Contact Intact Skin Only

General explanation of the rule: Rule 1 forms a fallback rule that is applicable to all medical devices that are not generally covered by a more specific rule later. It applies in general to all devices that meet only intact skin, and which does not touch the patient (Table 8.5).

Table 8.5: Classification Rules as per Rule 1 of MEDDEV 2.4/1

Rule 1	Examples
All non-invasive devices are in Class I, unless one of the rules set out hereinafter applies.	• Body liquid collection devices intended to be used in such a way that a return flow is unlikely (e.g., to collect body wastes such as urine collection bottles, ostomy pouches, incontinence pads or collectors used with wound drainage devices). They may be connected to the patient by means of catheters and tubing. • Devices used to immobilize body parts and/or to apply force or compression on them (e.g., non-sterile dressings used to aid the healing of a sprain, plaster of Paris, cervical collars, gravity traction devices, compression hosiery). • Devices intended in general for external patient support (e.g., hospital beds, patient hoists, walking aids, wheelchairs, stretchers, dental patient chairs). • Corrective glasses, frames, stethoscopes for diagnosis, eye occlusion plasters, incision drapes, conductive gels, non-invasive electrodes (electrodes for EEG or ECG), image intensifying screens. • Permanent magnets for removal of ocular debris

8.4.1.1. Understanding the Practical Issues Around Rule 1 Classification

Some non-invasive devices are indirectly in contact with the human body and can influence internal physiological processes by storing, channeling, or treating blood, other body liquids or liquids which are returned or infused into the body or by generating energy that is delivered to the body. These must be excluded when applying this rule and should be handled by another rule because of the hazards inherent in such direct influence on the body.

8.4.2. Rule 2: Channeling or Storing for Eventual Administration

General explanation of the rule: these types of devices must be considered separately from the non-contact devices of Rule 1 because they may be indirectly invasive. The channel or store substances that will be eventually delivered into the body. Typically, these devices after used in transfusion, infusion, extracorporeal circulation, delivery of anesthetic gases and oxygen. In some cases, devices covered under this rule are very simple devices that are gravity activated (Table 8.6).

Table 8.6: Classification Rules as per Rule 2 of MEDDEV 2.4/1

Rule 2	Examples
All non-invasive devices intended for channeling or storing blood, body-liquids or tissues, liquids, or gases for the purpose of eventual infusion, administration or introduction into the body are in Class IIa: • If they may be connected to an active medical device in Class IIa or a higher class	• Devices intended to be used as channels inactive drug delivery systems, e.g., tubing intended for use with an infusion pump. • Devices used for channeling, e.g., antistatic tubing for anesthesia, anesthesia breathing circuits and pressure indicator, pressure limiting devices. • Syringes for infusion pumps.
If they are intended for use for storing or channeling blood or other body liquids or for storing organs, parts of organs or body tissues (are in Class II-a)	• Devices intended to channel blood (e.g., in transfusion, extracorporeal circulation). • Devices intended for temporary storage and transport of organs for transplantation. • Devices intended for long term storage of biological substances and tissues such as corneas, sperm, human embryos, etc.
In all other cases they are in Class I	• Devices that provide a simple channeling function, with gravity providing the force to transport the liquid, e.g., administration sets for infusion. • Devices intended to be used for a temporary containment or storage function such as cups and spoons specifically intended for administering medicines. • Syringes without needle

8.4.2.1. Understanding the Practical Issues Around Rule 2 Classification

If a device can be used for a purpose that would result in it to be connected to an active device, then such a device will automatically fall in Class II A unless the manufacturer clearly states that it should not be connected to an active device of Class II-A or higher. It should also be noted that within Rule 2, blood bags are covered as an exception under a separate rule (Rule 18).

Special concepts are also noted in Table 8.6 for Rule 2 and are explained as follows:

- **Note 1:** All non-invasive devices intended for channeling or storing blood, body-liquids or tissues, liquids, or gases for the purpose of eventual infusion, administration or introduction into the body are in Class IIa: *If they may be connected to an active medical device in Class IIa or a higher class.*

Such connection is deemed to exist between a non-active device and an active device where a non-active device forms a link in the transfer of the substance between the patient and the active device. Furthermore, the safety and performance of one of the devices is influenced by the other device.

An example as explained in the guideline, would be tubing in an extracorporeal circulation system which is downstream from a blood pump and in the same blood flow circuit but not directly in contact with the pump.

- **Note 2:** "Devices intended to be used for a temporary containment or storage function such as cups and spoons specifically intended for administering medicines."

This statement refers to solutions intended for the preservation of organs during storage and transport and not medical devices.

8.4.3. Rule 3: Devices That Modify Biological or Chemical Composition of Blood, Body Liquids or Other Liquids

General explanation of the rule: these types of devices must be considered separately from non-contact devices of Rule 1 because they are indirectly invasive and treat or modify substances that will eventually be delivered into the body. Rule 3 covers mostly the more sophisticated elements of extracorporeal circulation sets, dialysis systems and autotransfusion systems as well as devices for extracorporeal treatment of body fluids which may not be reintroduced immediately into the body, including where the patient is not in a closed loop with the device (Table 8.7).

Table 8.7: Classification Rules as per Rule 3 of MEDDEV 2.4/1

Rule 3	Examples
All non-invasive devices intended for modifying the biological or chemical composition of blood, other body liquids or other liquids intended for infusion into the body are in Class IIb	• Devices intended to remove undesirable substances out of the blood by exchange of solutes such as hemodialyzers. • Devices intended to separate cells by physical means, e.g., gradient medium for sperm separation. • Hemodialysis concentrates
• Unless the treatment consists of filtration, centrifugation or exchange of gas or heat, in which case they are in Class IIa	–
• Unless the treatment consists of filtration, centrifugation or exchange of gas or heat, in which case they are in Class IIa	• Particulate filtration of blood in an extracorporeal circulation system. These are used to remove particles and emboli from the blood. • Centrifugation of blood to prepare it for transfusion or autotransfusion. -Removal of carbon dioxide from the blood and/or adding oxygen. • Warming or cooling the blood in an extracorporeal circulation system

8.4.3.1. Understanding the Practical Issues of Rule 3 Classification

These devices are normally used in conjunction with an active medical device covered under Rule 9 or Rule 11. Filtration and centrifugation should be understood in the context of this rule as exclusively mechanical methods.

8.4.4. Rule 4: Devices in Contact with Injured Skin

General explanation of the rule: Rule 4 is intended to cover mainly wound dressings regardless/independent of the wound. Traditional types of products such as those commonly used as mechanical barriers, are often well understood and do not result in any great hazard. In addition, there have been rapid technological developments in this area, with the emergence of new types of wound dressings for which non-traditional claims are made.

Examples of non-traditional claims include management of the micro-environment of a wound to enhance its natural healing mechanisms. Other ambitious claims may also be included, and these involve the mechanism of healing by secondary intent. This rule also includes some devices used on breached dermis which may even have a life-sustaining or lifesaving purpose, such as when there is full-thickness destruction of the skin over a large area and/or systemic effect. Furthermore, dressings containing medicinal products acting as ancillary to the dressing fall within Class III under Rule 13 (Table 8.8).

Table 8.8: Classification Rules as per Rule 4 of MEDDEV 2.4/1

Rule 4	**Examples**
All non-invasive devices which come into contact with injured skin: • are in Class I if they are intended to be used as a mechanical barrier, for compression or for absorption of exudates,	• Wound dressings, such as absorbent pads, island dressings, cotton wool, wound strips and gauze dressings to act as a barrier or to maintain the wound positionally or to absorb exudates from the wound.
• are in Class IIb if they are intended to be used principally with wounds which have breached the dermis and can only heal by secondary intent	• Are principally intended to be used with severe wounds that have substantially and extensively breached the dermis, and where the healing process can only be by secondary intent such as: o dressings for chronic extensive ulcerated wounds o dressings for severe burns having breached the dermis and covering an extensive area o dressings for severe wounds o dressings incorporating means of augmenting tissue and providing a temporary skin substitute
• are in Class IIa in all other cases, including devices principally intended to manage the micro-environment of a wound.	• Have specific properties intended to assist the healing process by controlling the level of moisture at the wound during the healing process and to generally regulate the environment in terms of humidity and temperature, levels of oxygen and other gases and pH values or by influencing the process by other physical means. • These devices may specify additional healing properties whilst not being intended for extensive wounds requiring healing by secondary intent. • Adhesives for topical use. • Polymer film dressings, hydrogel dressings, and non-medicated impregnated gauze dressings.

8.4.4.1. Understanding the Practical Issues of Rule 4 Classification

The products that fall within this rule are extremely claim sensitive, for example, a polymeric film dressing would fall in Class II A if the intended use is to manage the micro-environment of the wound and in Class I if its intended use is limited to retaining an invasive cannula at the wound site. Thus, the actual use of the device would determine the class, and it is impossible to say that a dressing is in a given class without knowing its intended use as defined by the manufacturer. So, for example, a claim that the device is interactive or active with respect to the wound healing process usually implies that the device is in Class II B.

It should also be noted that most dressings intended for a function/use that is in Class II A or II B will also perform functions that are in Class I, and as a result, such devices are nevertheless classed according to the intended use in the higher class.

For such devices incorporating medicines, Rule 13 should be consulted, and for devices incorporating animal tissues, Rule 17 should be consulted.

Special concepts defined by this rule include:

1. **Breached Dermis:** Whereby the wound exposes at least partly the subcutaneous tissues.
2. **Secondary Intent:** whereby the wound heals by first being filled with granulation tissue. This is followed by the epithelium, which grows back over the granulation tissue and the wound contracts. This is in contrast to primary intent, which implies that the edges of the wound are close enough or pulled together, e.g., by suturing, to allow the wound to heal.

8.4.5. Rule 5: Devices Invasive in Body Orifices

General explanation of the rule: Rule 5 is intended to cover invasiveness with respect to the body orifices such as the ear, mouth, nose, eye, anus, urethra, and vagina. This must be considered separately from invasiveness that penetrates through a cut in the body surfaces, for example through surgical invasiveness. A further distinction must be made between invasiveness with respect to the less vulnerable anterior parts of the ear, mouth, and nose and the other anatomical sites that can be accessed through natural body orifices. The surgically created stoma, which for example allows the evacuation of urine or feces, should also be considered as a body orifice. Devices covered by this rule tend to be diagnostic and therapeutic instruments used in particular specialties (ENT, ophthalmology, dentistry, proctology, urology, and gynecology) (Table 8.9).

Table 8.9: Classification Rules as per Rule 5 of MEDDEV 2.4/1

Rule 5	Examples
All invasive devices with respect to body orifices, other than surgically invasive devices and which are not intended for connection to an active medical device: • are in Class I if they are intended for transient use • are in Class IIa if they are intended for short term use except if they are used in the oral cavity as far as the pharynx, in an ear canal up to the eardrum or in a nasal cavity, in which case they are in Class I: • are in Class IIb if they are intended for long term use except if they are used in the oral cavity as far as the pharynx, in an ear canal up to the eardrum or in a nasal cavity and are not liable to be absorbed by the mucous membrane, in which case they are in Class IIa All invasive devices with respect to body orifices, other than surgically invasive devices, intended for connection to an active medical device in Class IIa or a higher class, are in Class IIa.	• Handheld mirrors used in dentistry to aid in dental diagnosis and surgery, dental impression materials, tubes used for pumping the stomach, impression tray, enema devices, examination gloves, and prostatic balloon dilation catheters. • Contact lenses, urinary catheters, tracheal tubes, stents, vaginal pessaries, and perineal re-education devices. • Dressings for nose bleeds, dentures intended to be removed by the patient • Urethral stents • Orthodontic wire, fixed dental prostheses, fissures sealants. • Tracheostomy or tracheal tubes connected to a ventilator, blood oxygen analyzers placed under the eyelid, powered nasal irrigators, nasopharyngeal airways, some enteral feeding tubes, fiber optics in endoscopes connected to surgical lasers, suction catheters or tubes for stomach drainage, dental aspirator tips.

8.4.6. Rule 6: Surgically Invasive Devices for Transient use

General explanation of the rule: This rule covers principally three major groups of devices, namely 1) devices that are used to create a conduit through the skin such as needles; 2) surgical instruments such as scalpels and saws; 3) various types of catheters and/or suckers and so on (Table 8.10).

Table 8.10: Classification Rules as per Rule 6 of MEDDEV 2.4/1

Rule 6	Examples
All surgically invasive devices intended for transient use are in Class IIa unless they are: • intended specifically to diagnose, monitor, or correct a defect of the heart or of the central circulatory system through direct contact with these parts of the body, in which case they are in Class III • reusable surgical instruments, in which case they are in Class I • intended to supply energy in the form of ionizing radiation in which case they are in Class IIb • intended to have a biological effect or to be wholly or mainly absorbed in which case they are in Class IIb • intended to administer medicines by means of a delivery system, if this is done in a manner that is potentially hazardous considering of the mode of application, in which case they are Class IIb.	• Needles used for suturing, needles or syringes, lancets, suckers, single-use scalpels, single-use scalpel blades, support devices in ophthalmic surgery, staplers, surgical swabs, drill bits connected to active devices, surgical gloves, etchants, tester of artificial heart valves, heart valve sizers and holders, trial hip prosthesis heads or stems, swabs to sample exudates, single-use aortic punches (see Note 2) • Cardiovascular catheters (e.g., angioplasty balloon catheters), including related guidewires and dedicated disposable cardiovascular surgical instruments, e.g., electrophysiological catheters, electrodes for electrophysiological diagnosis and ablation. -Catheters containing or incorporating sealed radioisotopes, where the radioactive isotope as such is not intended to be released into the body, if used in the central circulatory system • Scalpels, scalpel handles, drill bits, saws, that are not intended for connection to an active device, and retractors forceps, excavators, and chisels. • Catheters containing or incorporating sealed radioisotopes, where the radioactive isotope as such is not intended to be released into the body, if used in the circulatory system, excluding the central circulatory system • Devices for repeated self-application where dosage levels and the nature of the medicinal product are critical, e.g., insulin pens.

8.4.6.1. Explanation of Special Concepts Applicable in Rule 6

- **Note 1:** Terms such as *"surgically invasive device," "central circulatory system"* and *"reusable surgical instruments"* are defined in Section I of Annex IX of the Directive. Surgical instruments that are connected to an active device are not considered to be "reusable surgical instruments."
- **Note 2:** The phrase "correct a defect" does not cover devices that are used as an accessory in heart surgery, for example clamps. The first indent of this rule does not apply to aortic punches and similar cutting instruments which perform a similar function to that of a scalpel.

- **Note 3:** This applies to surgical instruments which are not specifically intended for purposes described in the first indent of the table, and irrespective of the site of application, are 1) in Class IIA if they are intended for single-use and 2) in Class I if they are reusable.
- **Note 4:** Dedicated means that the intended purpose of the device is to diagnose, monitor or correct a defect of the heart or of the central circulatory system.
- **Note 5:** The phrase "biological effect" pertains to all materials and devices that have the potential to affect tissues following use in a surgically invasive procedure. A material is considered to have some kind of biological effect if it actively and intentionally induces, alters, or prevents a response from the tissues that is mediated by specific reactions at a molecular level. These kinds of devices may also be described as bioactive.

Wholly or mainly absorbed: Th term absorption refers to the degradation of a material within the body and the metabolic elimination of the resulting degradation products from the body.

- **Note 6:** The concept of *"potentially hazardous manner"* is related to the characteristics of the device and not the competence of the user.

8.4.7. Rule 7: Surgically Invasive Devices for Short-Term Us

General explanation of the rule: Devices covered by Rule 7 are most those devices used in the context of surgery or post-operative care such as clamps or drains; infusion devise such as cannula or needles and catheters of various types (Table 8.11).

Table 8.11: Classification Rules as per Rule 7 of MEDDEV 2.4/1

Rule 7	**Examples**
All surgically invasive devices intended for short term use are in Class IIa unless they are intended: • either specifically to diagnose, monitor or correct a defect of the heart or of the central circulatory system through direct contact with these parts of the body, in which case they are in Class III;	• Clamps, infusion cannula, skin closure devices, temporary filling materials. • Tissue stabilizers used in cardiac surgery • Cardiovascular catheters, cardiac output probes and temporary pacemaker leads. • Thoracic catheters intended to drain the heart, including the pericardium • Carotid artery shunts

• or specifically for use in direct contact with the central nervous system, in which case they are in Class III; • or to supply energy in the form of ionizing radiation in which case they are in Class IIb • intended to have a biological effect or to be wholly or mainly absorbed in which case they are in Class III; • or to undergo chemical change in the body, except if the devices are placed in the teeth, or to administer medicines, in which case they are Class IIb.	• Neurological catheters, cortical electrodes • Brachytherapy devices • Absorbable sutures and biological adhesives • Adhesives

8.4.7.1. Practical Issues of Rule 7 Classification

- **Note 1:** *"administer medicines"*—the administration of medicines is more than just channeling as it implies also storage and/or influencing the volume and rate of medicine being delivered. It should also be noted that implanted capsules for the slow release of medicines are actually medicines and not medical devices.
- **Note 2:** *"correct a defect"*—this phrase does not cover devices that are used as an accessory in heart surgery, for example tissue stabilizers.

8.4.8. Rule 8: Surgically Invasive Devices for Long-Term Use and Implantable Devices

General explanation of the rule: devices falling under Rule 8 are mostly implants in the orthopedic, dental, ophthalmic, and cardiovascular fields, as well as soft tissue implants such as implants used in plastic surgery (Table 8.12).

Table 8.12: Classification Rules as per Rule 8 of MEDDEV 2.4/1

Rule 8	**Examples**
All implantable devices and long-term surgically invasive devices are in Class IIb unless they are intended: • to be placed in the teeth, in which case they are in Class IIa	• Prosthetic joint replacements, ligaments, shunts, stents, nails, plates, intra-ocular lenses, internal closure devices, tissue augmentation implants, infusion ports, peripheral vascular grafts, penile implants,

• to be used in direct contact with the heart, the central circulatory system, or the central nervous system, in which case they are Class III; • to have a biological effect or to be wholly or mainly absorbed, in which case they are in Class III; • or to undergo chemical change in the body, except if the devices are placed in the teeth, or to administer medicines, in which case they are in Class III.	non-absorbable sutures, bone cements and maxillo-facial implants, viscoelastic surgical devices intended specifically for ophthalmic anterior segment surgery • Bridges, crowns, dental filling materials and pins, dental alloys, ceramics, and polymers. • Prosthetic heart valves, aneurysm clips, vascular prostheses, spinal stents, vascular stents, CNS electrodes and cardiovascular sutures.

8.4.8.1. Practical Issues of Classification for Rule 8

- **Note 1:** These products are implants because in normal conditions, a significant amount of the substance remains at the surgical site after the procedure. If, however, these devices contain animal tissues or derivatives of animal tissues, they are covered by Rule 17.
- **Note 2:** "to be placed in the teeth"—implants without bioactive coatings intended to secure teeth or prostheses to the maxillary or mandibular bones are in Class II B following the general rule.
- **Note 3:** "undergo chemical change"—the clause about chemical change under this rule does not apply to products such as bone cements where the chemical change takes place during the placement and does not continue in the long term.

8.4.9. Rule 9: Active Therapeutic Devices Intended to Administer or Exchange Energy

General explanation of the rule: Medical devices classified by Rule 8 are mostly electrical equipment used in surgery, such as lasers and surgical generators. There are, however, devices for specialized treatment such as radiation treatment. In addition, another category consists of stimulation devices, although not all of them can be considered as delivering dangerous levels of energy considering the tissue involved (Table 8.13).

Table 8.13: Classification Rules as per Rule 9 of MEDDEV 2.4/1

Rule 9	**Examples**
All active therapeutic devices intended to administer, or exchange energy are in Class IIa Unless their characteristics are such that they may administer or exchange energy to and from the human body in a potentially hazardous way, taking account of the nature, the density, and the site of application of the energy, in which case they are in Class IIb.	• Electrical and/or magnetic and electromagnetic energy: Muscle stimulators and external bone growth stimulators, TENS devices and eye electromagnets, electrical acupuncture • Thermal energy: Cryosurgery equipment, heat exchangers, except the types described below. • Mechanical energy: Powered dermatomes, powered drills, and dental handpieces. • Light: Phototherapy for skin treatment and for neonatal care • Sound: Hearing aids • Ultrasound: Equipment for physiotherapy • Kinetic energy-lung ventilators • Thermal energy-Incubators for babies, warming blankets, blood warmers, electrically powered heat exchangers, for instance those used with patient's incapable of reacting, communicating, and/or who are without a sense of feeling • Electrical energy-high-frequency electrosurgical generators, and electrocautery equipment, including their electrodes, external pacemakers, external defibrillators, electroconvulsive therapy equipment • Coherent light-surgical lasers • Ultrasound-surgical ultrasound devices • Ionizing radiation-radioactive sources for afterloading therapy, therapeutic cyclotrons, linear accelerators, therapeutic X-ray sources.
All active devices intended to control and monitor the performance of active therapeutic devices in Class IIb or intended to influence directly the performance of such devices are in Class IIb.	• External feedback systems for active therapeutic devices, after loading control devices

8.4.9.1. Explanation of Special Concepts Pertaining to Rule 9

- **Note 1:** *"exchange energy to and from the human body in a potentially hazardous way"*

The decision as to whether a medical device administers or exchanges energy to and from the human body in a potentially hazardous way should take various factors into consideration. The concept of "potentially hazardous" is dependent on the type of technology involved and the intended application of the device to the patient, and not on the measures adopted by the manufacturer in view of good design management. That being said, the manufacturer's obligation to comply with design requirements and solutions adopted (such as the use of standards) exist independently from the classification system.

8.4.10. Rule 10: Active Devices for Diagnosis

A general explanation of the rule: this rule covers primarily a whole range of widely used equipment in the field's ultrasound diagnosis and the capture of physiological signals as well as therapeutic and diagnostic radiology (Table 8.14).

Table 8.14: Classification Rules as per Rule 10 of MEDDEV 2.4/1

Rule 10	Examples
Active devices intended for diagnosis are in Class IIa: • if they are intended to supply energy which will be absorbed by the human body, except for devices used to illuminate the patient's body, in the visible spectrum; • if they are intended to image *in vivo* distribution of radiopharmaceuticals; • if they are intended to allow direct diagnosis or monitoring of vital physiological processes Unless they are specifically intended for monitoring of vital physiological parameters, where the nature of variations is such that it could result in immediate danger to the patient, for instance, variations in cardiac performance, respiration, activity of CNS in which case they are in Class IIb. Active devices intended to emit ionizing radiation and intended for diagnostic and therapeutic interventional radiology, including devices which control or monitor such devices, or which directly influence their performance, are in Class II B.	• Magnetic resonance equipment, pulp testers, evoked response stimulators, diagnostic ultrasound • Gamma cameras, positron emission tomography and single-photon emission computer tomography • Electrocardiographs, electroencephalographs, cardioscopes with or without pacing pulse indicators • Electronic thermometers • Electronic stethoscopes • Electronic blood pressure measuring equipment • Intensive care monitoring and alarm devices (for example, blood pressure, temperature, oxygen saturation), biological sensors, blood gas analyzers used in open-heart surgery, cardioscopes, and apnea monitors, including apnea monitors in-home care • Diagnostic X-ray sources.

8.4.10.1. Explanation of Special Concepts Pertaining to Rule 10

- **Note 1:** "vital physiological processes"—these vital physiological processes and parameters include for example; respiration, heart rate, cerebral functions, blood gases, blood pressure, and body temperature. Medical devices intended to be used for continuous surveillance of vital physiological processes in anesthesia, intensive care or emergency care are in Class IIB, whilst medical devices intended to be used to obtain readings of vital physiological signals in routine check-ups and in self-monitoring are in Class IIA. A thermal imaging device intended to monitor blood flow is not considered to be a temperature measuring device.
- **Note 2:** "With or without pacing pulse indicators"—this is applicable for devices specifically intended to monitor AIMDs and which fall under the AIMD Directive.
- **Note 3:** Therapeutic interventional radiology. This refers to diagnosis being carried out during surgical procedures.
- **Note 4:** "including devices which control or monitor"—this refers to active devices for the control, monitoring or influencing and not to the subsequent processing, recording, or viewing of the resulting image.

8.4.11. Rule 11: Active Devices to Administer, Remove Medicines and Other Substances to or From the Body

General explanation of the rule: this rule is intended to cover primarily drug delivery systems and anesthesia equipment (Table 8.15).

Table 8.15: Classification Rules as per Rule 11 of MEDDEV 2.4/1

Rule 11	**Examples**
All active devices intended to administer and/or remove medicines, body-liquids or other substances to or from the body are in Class IIa. Unless this is done in a manner: -that is potentially hazardous, taking account of the nature of the substances involved, of the part of the body concerned and of the mode of application, in which case they are in Class IIb.	• Suction equipment, feeding pumps. -Jet injectors for vaccination-Nebulizers to be used on conscious and spontaneously breathing patients where failure to deliver the appropriate dosage characteristics is not potentially hazardous

	• Infusion pumps, ventilators, anesthesia machines, anesthetic vaporizers, dialysis equipment, blood pumps for heart-lung machines, hyperbaric chambers, pressure regulators for medical gases, medical gas mixers, moisture exchangers in breathing circuits if used on unconscious or non-spontaneously breathing patients • Nebulizers where the failure to deliver the appropriate dosage characteristics could be hazardous.

8.4.12. Rule 12: All Other Active Devices

General explanation of the rule: this forms a fallback rule to cover all active devices not covered by previous rules (Table 8.16).

Table 8.16: Classification Rules as per Rule 12 of MEDDEV 2.4/1

Rule 12	Examples
All other active devices are in Class I	• Active diagnostic devices intended to illuminate the patient's body in the visible spectrum such as examination lights or to optically view the body such as surgical microscopes. • Devices intended in general for external patient support (e.g., hospital beds, patient hoists, walking aids, wheelchairs, stretchers, dental patient chairs). • Active diagnostic devices intended for thermography. • Active devices intended for recording, processing, or viewing of diagnostic images. • dental curing lights.

8.5. SPECIAL RULES

8.5.1. Rule 13: Devices Incorporating a Medicinal Substance

General explanation of the rule: This rule is intended to cover combination devices that contain a medicinal substance incorporated into the device. This substance is usually present for the purpose of assisting the functioning of that device. This rule, however, does not cover those devices incorporating substances which under other circumstances may be considered as medicinal substances. It does however cover that which is incorporated into the device exclusively for the purpose of maintaining certain characteristics of the device and which are not liable to act on the body. A common example includes agents for the preservation of solutions for contact lenses. The primary function of the device does not rely on the pharmacological effect of the medicine as if the latter is true, then the product would be classified as being a medicine rather than a device. In that case it will not be covered by the MEDDEV 2.4/1 Directive (Table 8.17).

Table 8.17: Classification Rules as per Rule 13 of MEDDEV 2.4/1

Rule 13	**Examples**
All devices incorporating, as an integral part, a substance which, if used separately, can be considered to be a medicinal product as defined in Article 1 of the Directive 65/65/EEC, and which is liable to act on the human body with action ancillary to that of the devices, are in Class III.	• Antibiotic bone cements, condoms with spermicide, heparin coated catheters, endodontic materials with antibiotics. • Ophthalmic irrigation solutions principally intended for irrigation, which contain components which support the metabolism of the endothelial cells of the cornea-Dressings incorporating an antimicrobial agent where the purpose of such an agent is to provide ancillary action on the wound

- **Note 1: Integral Part:** It means that the device and the medicinal substance forms one physical unit.

8.5.2. Rule 14: Devices Used for Contraception or Prevention of Sexually Transmitted Diseases

General explanation of the rule: the intended uses defined by this rule relates to special cases of human vulnerability that cannot be covered by the normal criteria of time, invasiveness, and organic function. While the rule does cover two very different device applications, some devices may perform both functions. A common example of this would be condoms. It should also be noted that devices intended to prevent the sexual transmission of HIV are also covered by this rule (Table 8.18).

Table 8.18: Classification Rules as per Rule 14 of MEDDEV 2.4/1

Rule 14	Examples
All devices used for contraception or the prevention of the transmission of sexually transmitted diseases are in Class IIb Unless they are implantable or long-term invasive devices, in which case they are in Class III.	• Condoms, contraceptive diaphragms • Contraceptive intrauterine devices (IUDs)

- **Note 1: Contraceptive Intrauterine Devices (IUDs):** Those intrauterine contraceptives whose primary purpose is to release progestogens are not medical devices.

8.5.3. Rule 15: Specific Disinfecting, Cleaning, and Rinsing Devices

General explanation of the rule: this rule is mainly intended to cover various contact lens fluids as well as those substances used principally in a medical environment to disinfect medical devices (Table 8.19).

Table 8.19: Classification Rules as per Rule 15 of MEDDEV 2.4/1

Rule 15	Examples
All devices intended specifically to be used for disinfecting, cleaning, rinsing, or, when appropriate hydrating contact lenses are in Class IIb. All devices intended specifically to be used for disinfecting medical devices are in Class IIa. This rule does not apply to products that are intended to clean medical devices other than contact lenses by means of physical action	• Contact lens solutions, comfort solutions. • Disinfectants specifically intended for instance for endoscopes or hemodialysis equipment, sterilizers specifically intended to sterilize medical devices in a medical environment and washer-disinfectors. • Cleaners which disinfect prosthetic dentures.

8.5.3.1. Practical Issues of Rule 15 Classification

- **Note 1:** This rule does not apply to mechanical means of cleaning of devices, such as brushes and ultrasound. Such products will only fall under this directive if they are specifically intended for use with medical devices.

8.5.4. Rule 16: Non-Active Devices to Record X-Ray Diagnostic Images (Table 8.20)

Table 8.20: Classification Rules as per Rule 16 of MEDDEV 2.4/1

Rule 16	Examples
Non-active devices specifically intended for recording of X-ray diagnostic images are in Class II a.	X-ray films, photostimulable phosphor plates

- **Note:** This rule refers to primary recording media such as X-ray films and not to media used for subsequent reproduction.

8.5.5. Rule 17: Devices Utilizing Animal Tissue or Derivatives

A general explanation of the rule: Rule 17 covers those devices that contain animal tissues, or which are made of animal tissues that have been rendered non-viable. It also considers those derivatives from such tissues where there is no longer any capacity for cellular metabolic activity (i.e.: non-viable tissues). It should be noted that those medical devices containing non-inactivated animal tissues and/or any human tissues or derivatives are excluded from the scope of this Directive. It should be further noted that the manufacture of some medical devices may use industrial raw materials, and this may result in the presence of small amounts of tallow or tallow derivatives which are not considered as derivatives of animal tissues for the purpose of this rule. In this case, this rule does not apply (Table 8.21).

Table 8.21: Classification Rules as per Rule 17 of MEDDEV 2.4/1

Rule 17	Examples
All devices manufactured utilizing animal tissues or derivatives rendered non-viable are Class III except where such devices are intended to come into contact with intact skin only	Biological heart valves, porcine xenograft dressings, catgut sutures, implants, and dressings made from collagen.

Practical classification issues: those medical devices made of non-viable animal tissue that come into contact with intact skin only are in Class I in accordance to Rule 1.

- **Note 1:** Animal tissues or derivatives—derivatives are defined as those products that are processed from animal tissues and exclude substances such as milk, silk, beeswax, hair, lanolin.

- **Note 2:** Intact skin includes the skin around an established stoma unless however, this skin is breached.

8.5.6. Rule 18: Blood Bags

A general explanation of the rule: This is a special rule that covers only blood bags (Table 8.22).

Table 8.22: Classification Rules as per Rule 18 of MEDDEV 2.4/1

Rule 18	Examples
By derogation from other rules, blood bags are in Class IIb	Blood bags (including those containing or coated with an anti-coagulant). Where blood bags have a function greater than for storing purposes and include systems for preservation other than anti-coagulants then other rules (e.g., Rule 13) may apply.

Chapter 9

Understanding EU Classification Rules under the MDR 2017/745

CONTENTS

9.1. UNDERSTANDING CLASSIFICATION RULES UNDER THE NEW MDR 2017/745

One of the major changes under the EU MDR that medical device manufacturers will see as they transition over to the new regulation is the change in requirements for device classification. It is also important to remember that as these medical device classifications change, so do the requirements imposed on manufacturers and their related interaction with their Notified Body.

Both the MDD and now the EU MDR 2017/745 classify medical devices into four classes (I, IIa, IIb, and III) based on their intended purposes and inherent risks. The higher risk of the medical device, the more rules, and regulations it must comply with.

9.2. UNDERSTANDING HOW CATEGORIES OF MEDICAL DEVICES ARE DEFINED UNDER MDR?

It is important to note that the duration of use and definition of the different product categories are very similar. There have, however, been some changes in vocabulary to further refine the understanding of such terminology. This similarity between the MDD and MDR is important as if there was such a definition change, it would most likely mean a manufacturer has to reconsider their product's risk classification and take into consideration the related repercussions on the business.

- **Chapter I:** Understanding the definitions specific to classification rules as defined in the MDR 2017/745.

Chapter I of the MDR 2017/745: Definitions Specific to Classification Rules
Duration of use: • 'Transient' means normally intended for continuous use for less than 60 minutes. • 'Short term' means normally intended for continuous use for between 60 minutes and 30 days. 1. • 'Long term' means normally intended for continuous use for more than 30 days.
Invasive and active devices: • 'Body orifice' means any natural opening in the body, as well as the external surface of the eyeball, or any permanent artificial opening, such as a stoma. • 'Surgically invasive device' means: o an invasive device which penetrates inside the body through the surface of the body, including through mucous membranes of body orifices with the aid or in the context of a surgical operation; and

o A device which produces penetration other than through a body orifice.
'Reusable surgical instrument' means an instrument intended for surgical use in cutting, drilling, sawing, scratching, scraping, clamping, retracting, clipping or similar procedures, without a connection to an active device and which is intended by the manufacturer to be reused after appropriate procedures such as cleaning, disinfection, and sterilization have been carried out.
'Active therapeutic device' means any active device used, whether alone or in combination with other devices, to support, modify, replace, or restore biological functions or structures with a view to treatment or alleviation of an illness, injury, or disability.
'Active device intended for diagnosis and monitoring' means any active device used, whether alone or in combination with other devices, to supply information for detecting, diagnosing, monitoring, or treating physiological conditions, states of health, illnesses, or congenital deformities.
'Central circulatory system' means the following blood vessels: arteriae pulmonales, aorta ascendens, arcus aortae, aorta descendens to the bifurcation aortae, arteriae coronariae, arteria carotiscommunis, arteria carotis externa, arteria carotisinterna, arteriae cerebrales, truncus brachiocephalic, venae cordis, venae pulmonales, vena cava superior and vena cava inferior.
'Central nervous system' means the brain, meninges, and spinal cord.
'Injured skin or mucous membrane' means an area of skin or a mucous membrane presenting a pathological change or change following disease or a wound.

What might also help a manufacturer is a simple comparison of definitions between the MDR and MDD. Viewing this side by side will make it easier for a manufacturer to decipher the changes between the two regulations (**Table 9.1**).

Table 9.1: Comparison of MDR and MDD Definitions

Term	MDR	MDD	Explanation of Changes
Surgically invasive device	An invasive device which penetrates inside the body through the surface of the body, including through mucous membranes of body orifices with the aid or in the context of a surgical operation; (Annex VIII)	An invasive device which penetrates inside the body through the surface of the body, with the aid or in the context of a surgical operation.	There is now clarification that surgically invasive also includes penetration through mucous membranes of body orifices
Injured skin or mucous membrane	Means an area of skin or a mucous membrane presenting a pathological change or change following disease or a wound.	n/a	Introduced definition for injured skin or mucous membrane

Active device	Covered under Article 2. 'Active device' means any device, the operation of which depends on a source of energy other than that generated by the human body for that purpose, or by gravity, and which acts by changing the density of or converting that energy. Devices intended to transmit energy, substances, or other elements between an active device and the patient, without any significant change, shall not be deemed to be active devices. Software shall also be deemed to be an active device;	Any medical device operation of which depends on a source of electrical energy or any source of power other than that directly generated by the human body or gravity and which acts by converting this energy. Medical devices intended to transmit energy, substances, or other elements between an active medical device and the patient, without any significant change, is not considered to be active medical devices. Standalone software is considered to be an active medical device.	Device which acts by changing the density of energy are also considered active devices. This is considered a clarification only
Reusable surgical instrument	Instrument intended for surgical use in cutting, drilling, sawing, scratching, scraping, clamping, retracting, clipping or similar procedures, without a connection to an active device and which is intended by the manufacturer to be reused after appropriate procedures such as cleaning, disinfection, and sterilization have been carried out.	Instrument intended for surgical use by cutting, drilling, sawing, scratching, scraping, clamping, retracting, clipping or similar procedures, without connection to any active medical device and which can be reused after appropriate procedures have been carried out.	Devices which can be reused vs. intended by the manufacturer to be reused; mostly a clarification

Continuous use	(a) The entire duration of use of the same device without regard to temporary interruption of use during a procedure or temporary removal for purposes such as cleaning or disinfection of the device. Whether the interruption of use or the removal is temporary shall be established in relation to the duration of the use prior to and after the period when the use is interrupted or the device removed; and (b) the accumulated use of a device that is intended by the manufacturer to be replaced immediately with another of the same type.	Means 'an uninterrupted actual use of the device for the intended purpose.' However, where usage of a device is discontinued in order for the device to be replaced immediately by the same or an identical device, this shall be considered an extension of the continuous use of the device	Concept of continuous use extended to include devices that may be temporarily removed to be cleaned or disinfected and then re-used; However, this is still considered a clarification since such use would be treated as 'continuous use' under the Directive as well
Direct diagnosis	A device is considered to allow direct diagnosis when it provides the diagnosis of the disease or condition in question by itself or when it provides decisive information for the diagnosis.	n/a	New definition
Implantable device	Covered under Article 2.5 'implantable device' means any device, including those that are partially or wholly absorbed, which is intended: to be totally introduced into the human body, or—to replace an epithelial surface or the surface of the eye, by clinical intervention and which is intended to remain in place after the procedure. Any device intended to be partially introduced into the human body by clinical intervention and intended to remain in place after the procedure for at least 30 days shall also be deemed to be an implantable device;	Any device which is intended: —to be totally introduced into the human body or, —to replace an epithelial surface or the surface of the eye, by surgical intervention which is intended to remain in place after the procedure. Any device intended to be partially introduced into the human body through surgical intervention and intended to remain in place after the procedure for at least 30 days is also considered an implantable device.	Devices that are partially or wholly absorbed are considered implantable; 'Clinical' intervention vs. 'surgical' intervention

- **Chapter II:** Implementing classification rules are defined in Annex VIII of the MDR 2017/745 as follows:

Chapter II of the MDR 2017/745: Implementing Rules
Application of the classification rules shall be governed by the intended purpose of the devices. If the device in question is intended to be used in combination with another device, the classification rules shall apply separately to each of the devices. Accessories for a medical device and for a product listed in Annex XVI shall be classified in their own right separately from the device with which they are used. Software, which drives a device or influences the use of a device, shall fall within the same class as the device. If the software is independent of any other device, it shall be classified in its own right. If the device is not intended to be used solely or principally in a specific part of the body, it shall be considered and classified on the basis of the most critical specified use. If several rules, or if, within the same rule, several sub-rules, apply to the same device based on the device's intended purpose, the strictest rule and sub-rule resulting in the higher classification shall apply. In calculating the duration referred to in Section 1, continuous use shall mean: a. The entire duration of use of the same device without regard to temporary interruption of use during a procedure or temporary removal for purposes such as cleaning or disinfection of the device. Whether the interruption of use or the removal is temporary shall be established in relation to the duration of the use prior to and after the period when the use is interrupted or the device removed; and b. The accumulated use of a device that is intended by the manufacturer to be replaced immediately with another of the same type. A device is considered to allow direct diagnosis when it provides the diagnosis of the disease or condition in question by itself or when it provides decisive information for the diagnosis.

While the MDD allows for 18 classification rules, the new MDR contains 22 rules. These 18 rules can be found in Annex IX of the MDD. The MDR on the other hand contains 22 rules, of which the 18 classification rules of the MDD are still present. These 22 rules can be found in Annex VIII of the MDR.

There are, however, four more rules than the previous medical device directive (MDD), which is partly due to the broader scope of the MDR. This means the MDR will apply to certain products that are not regulated through the MDD. All of the rules are based on the potential risks associated with the device itself, its technical design as well as how the device itself is manufactured.

The 22 device classification rules are defined in Annex VIII of the MDR 2017/745 as follows:

Chapter III of the MDR 2017/745: Classification Rules

Non-Invasive Devices

- **Rule 1:** All non-invasive devices are classified as Class I, unless one of the rules set out hereinafter applies.

- **Rule 2:** All non-invasive devices intended for channeling or storing blood, body-liquids, cells or tissues, liquids, or gasses for the purpose of eventual infusion, administration or introduction into the body are classified as Class IIa:
  - If they may be connected to a Class IIa, Class IIb or Class III active device; or
  - If they are intended for use for channeling or storing blood or other body liquids or for storing organs, parts of organs or body cells and tissues, except for blood bags, blood bags are classified as Class IIb.

In all other cases, such devices are classified as Class I.

- **Rule 3:** All non-invasive devices intended for modifying the biological or chemical composition of human tissues or cells, blood, other body liquids or other liquids intended for implantation or administration into the body are classified as Class IIb, unless the treatment for which the device is used consists of filtration, centrifugation or exchanges of gas, heat, in which case they are classified as Class IIa.

All non-invasive devices consisting of a substance or a mixture of substances intended to be used *in vitro* in direct contact with human cells, tissues, or organs taken from the human body or used *in vitro* with human embryos before their implantation or administration into the body are classified as Class III.

- **Rule 4:** All non-invasive devices which come into contact with injured skin or mucous membrane are classified as:
  - Class I if they are intended to be used as a mechanical barrier, for compression or for absorption of exudates;
  - Class IIb if they are intended to be used principally for injuries to skin which have breached the dermis or mucous membrane and can only heal by secondary intent;
  - Class IIa if they are principally intended to manage the micro-environment of injured skin or mucous membrane; and
  - Class IIa in all other cases. This rule applies also to the invasive devices that come into contact with injured mucous membrane.

Invasive Devices

- **Rule 5:** All invasive devices with respect to body orifices, other than surgically invasive devices, which are not intended for connection to an active device or which are intended for connection to a Class I active device are classified as:
  - Class I if they are intended for transient use;
  - Class IIa if they are intended for short-term use, except if they are used in the oral cavity as far as the pharynx, in an ear canal up to the eardrum or in the nasal cavity, in which case they are classified as Class I; and
  - Class IIb if they are intended for long-term use, except if they are used in the oral cavity as far as the pharynx, in an ear canal up to the eardrum or in the nasal cavity and are not liable to be absorbed by the mucous membrane, in which case they are classified as Class IIa.

All invasive devices with respect to body orifices, other than surgically invasive devices, intended for connection to a Class IIa, Class IIb, or Class III active device, are classified as Class IIa.

➢ **Rule 6:** All surgically invasive devices intended for transient use are classified as Class IIa unless they: • Are intended specifically to control, diagnose, monitor, or correct a defect of the heart or of the central circulatory system through direct contact with those parts of the body, in which case they are classified as Class III; • Are reusable surgical instruments, in which case they are classified as Class I; • Are intended specifically for use in direct contact with the heart or central circulatory system or the central nervous system, in which case they are classified as Class III; • Are intended to supply energy in the form of ionizing radiation in which case they are classified as Class IIb; • Have a biological effect or are wholly or mainly absorbed in which case they are classified as Class IIb; or • Are intended to administer medicinal products by means of a delivery system, if such administration of a medicinal product is done in a manner that is potentially hazardous taking account of the mode of application, in which case they are classified as Class IIb.
• **Rule 7:** All surgically invasive devices intended for short-term use are classified as Class IIa unless they: • Are intended specifically to control, diagnose, monitor, or correct a defect of the heart or of the central circulatory system through direct contact with those parts of the body, in which case they are classified as Class III; • Are intended specifically for use in direct contact with the heart or central circulatory system or the central nervous system, in which case they are classified as Class III; • Are intended to supply energy in the form of ionizing radiation in which case they are classified as Class IIb; • Have a biological effect or are wholly or mainly absorbed in which case they are classified as Class III; • Are intended to undergo chemical change in the body in which case they are classified as Class IIb, except if the devices are placed in the teeth; or • Are intended to administer medicines, in which case they are classified as Class IIb.
• **Rule 8:** All implantable devices and long-term surgically invasive devices are classified as Class IIb unless they: • Are intended to be placed in the teeth, in which case they are classified as Class IIa; • Are intended to be used in direct contact with the heart, the central circulatory system, or the central nervous system, in which case they are classified as Class III; • Have a biological effect or are wholly or mainly absorbed, in which case they are classified as Class III; • Are intended to undergo chemical change in the body in which case they are classified as Class III, except if the devices are placed in the teeth; • Are intended to administer medicinal products, in which case they are classified as Class III; • Are active implantable devices or their accessories, in which cases they are classified as Class III; • Are breast implants or surgical meshes, in which cases they are classified as Class III; • Are total or partial joint replacements, in which case they are classified as Class III, with the exception of ancillary components such as screws, wedges, plates, and instruments; or • Are spinal disc replacement implants or are implantable devices that come into contact with the spinal column, in which case they are classified as Class III with the exception of components such as screws, wedges, plates, and instruments.

Active Devices
• **Rule 9:** All active therapeutic devices intended to administer or exchange energy are classified as Class IIa unless their characteristics are such that they may administer energy to or exchange energy with the human body in a potentially hazardous way, taking account of the nature, the density and site of application of the energy, in which case they are classified as Class IIb. All active devices intended to control or monitor the performance of active therapeutic Class IIb devices, or intended directly to influence the performance of such devices are classified as Class IIb. All active devices intended to emit ionizing radiation for therapeutic purposes, including devices that control or monitor such devices, or which directly influence their performance, are classified as Class IIb. All active devices that are intended for controlling, monitoring, or directly influencing the performance of active implantable devices are classified as Class III.
• **Rule 10:** Active devices intended for diagnosis and monitoring are classified as Class IIa: • If they are intended to supply energy which will be absorbed by the human body, except for devices intended to illuminate the patient's body, in the visible spectrum, in which case they are classified as Class I; • If they are intended to image *in vivo* distribution of radiopharmaceuticals; or • If they are intended to allow direct diagnosis or monitoring of vital physiological processes, unless they are specifically intended for monitoring of vital physiological parameters and the nature of variations of those parameters is such that it could result in immediate danger to the patient, for instance variations in cardiac performance, respiration, activity of the central nervous system, or they are intended for diagnosis in clinical situations where the patient is in immediate danger, in which cases they are classified as Class IIb. Active devices intended to emit ionizing radiation and intended for diagnostic or therapeutic radiology, including interventional radiology devices and devices which control or monitor such devices, or which directly influence their performance, are classified as Class IIb.
• **Rule 11:** Software intended to provide information which is used to make decisions with diagnosis or therapeutic purposes is classified as Class IIa, except if such decisions have an impact that may cause: • Death or an irreversible deterioration of a person's state of health, in which case it is in Class III; or • A serious deterioration of a person's state of health or a surgical intervention, in which case it is classified as Class IIb. Software intended to monitor physiological processes is classified as Class IIa, except if it is intended for monitoring of vital physiological parameters, where the nature of variations of those parameters is such that it could result in immediate danger to the patient, in which case it is classified as Class IIb. All other software is classified as Class I.
• **Rule 12:** All active devices intended to administer and/or remove medicinal products, body-liquids or other substances to or from the body are classified as Class IIa, unless this is done in a manner that is potentially hazardous, taking account of the nature of the substances involved, of the part of the body concerned and of the mode of application in which case they are classified as Class IIb.
• **Rule 13:** All other active devices are classified as Class I.
Special Rules

• **Rule 14:** All devices incorporating, as an integral part, a substance which, if used separately, can be considered to be a medicinal product, as defined in point 2 of Article 1 of Directive 2001/83/EC, including a medicinal product derived from human blood or human plasma, as defined in point 10 of Article 1 of that Directive, and that has an action ancillary to that of the devices, are classified as Class III.

• **Rule 15:** All devices used for contraception or prevention of the transmission of sexually transmitted diseases are classified as Class IIb, unless they are implantable or long-term invasive devices, in which case they are classified as Class III.

• **Rule 16:** All devices intended specifically to be used for disinfecting, cleaning, rinsing, or, where appropriate, hydrating contact lenses are classified as Class IIb.
All devices intended specifically to be used for disinfecting or sterilizing medical devices are classified as Class IIa, unless they are disinfecting solutions or washer-disinfectors intended specifically to be used for disinfecting invasive devices, as the endpoint of processing, in which case they are classified as Class IIb.
This rule does not apply to devices that are intended to clean devices other than contact lenses by means of physical action only.

• **Rule 17:** Devices specifically intended for recording of diagnostic images generated by X-ray radiation are classified as Class IIa.

• **Rule 18:** All devices manufactured utilizing tissues or cells of human or animal origin, or their derivatives, which are non-viable or rendered non-viable, are classified as Class III, unless such devices are manufactured utilizing tissues or cells of animal origin, or their derivatives, which are non-viable or rendered non-viable and are devices intended to come into contact with intact skin only.

• **Rule 19:** All devices incorporating or consisting of nanomaterial are classified as:
• Class III if they present a high or medium potential for internal exposure;
• Class IIb if they present a low potential for internal exposure; and
• Class IIa if they present a negligible potential for internal exposure.

• **Rule 20:** All invasive devices with respect to body orifices, other than surgically invasive devices, which are intended to administer medicinal products by inhalation are classified as Class IIa, unless their mode of action has an essential impact on the efficacy and safety of the administered medicinal product or they are intended to treat life-threatening conditions, in which case they are classified as Class IIb.

• **Rule 21:** Devices that are composed of substances or of combinations of substances that are intended to be introduced into the human body via a body orifice or applied to the skin and that are absorbed by or locally dispersed in the human body are classified as:
• Class III if they, or their products of metabolism, are systemically absorbed by the human body in order to achieve the intended purpose;
• Class III if they achieve their intended purpose in the stomach or lower gastrointestinal tract and they, or their products of metabolism, are systemically absorbed by the human body;
• Class IIa if they are applied to the skin or if they are applied in the nasal or oral cavity as far as the pharynx, and achieve their intended purpose on those cavities; and
• Class IIb in all other cases.

• **Rule 22:** Active therapeutic devices with an integrated or incorporated diagnostic function which significantly determines the patient management by the device, such as closed-loop systems or automated external defibrillators (AEDs), are classified as Class III.

The classification rules are summarized below:

Rule 1	Non-invasive devices.
Rule 2	Non-invasive devices intended for channeling or storing (Which includes cells).
Rule 3	Non-invasive devices that modify biological or chemical composition of blood, body-liquids, other liquids, and cells.
Rule 4	Non-invasive devices in contact with injured skin or mucous membrane.
Rule 5	Devices invasive in body orifices.
Rule 6	Surgically invasive devices for transient use.
Rule 7	Surgically invasive devices for short term use.
Rule 8	Surgically invasive devices for long term use and implantable (including any device administering medicinal products, surgical mesh, or spinal disc).
Rule 9	Active therapeutic devices intended to exchange or administer energy.
Rule 10	Active devices for diagnosis and monitoring, emit ionizing radiation.
Rule 11	Software intended to provide information which is used to make decisions with diagnosis or therapeutic purposes (from Class I to Class III).
Rule 12	Active devices intended to administer and/or remove medicinal products, body-liquids or other substances.
Rule 13	All other active devices.
Rule 14	Devices incorporating a medicinal substance including human blood or plasma.
Rule 15	Contraception or prevention of the transmission of sexually transmitted diseases.
Rule 16	Specific disinfecting, cleaning, and rinsing devices.
Rule 17	Devices specifically intended for recording of diagnostic images generated by X-ray radiation.
Rule 18	Devices utilizing non-viable tissues or cells of human origin or tissues of animal or derivatives.
Four New Rules Included in the MDR:	
Rule 19	Devices incorporating or consisting of nanomaterial.
Rule 20	Invasive devices with respect to body orifices to administer medicinal products by inhalation.
Rule 21	Substances or of combinations of substances that are intended to be introduced into the human body via a body orifice or applied to the skin and that are absorbed.
Rule 22	Active therapeutic devices with an integrated or incorporated diagnostic function which significantly determines the patient management.

The new regulation also comes with certain classification changes that are relevant to certain medical devices. All active implantable devices and their accessories will be considered as Class III. Any substance-based device intended to be used via a body orifice or applied on the skin may not be a

Class I. this means any substance-based device currently in Class I will be up-classified with the new regulation. This also means that a manufacturer with a device that will be up classified with the MDR must also comply with the stricter requirements arising from the new regulation.

9.3. COMPARISON OF CLASSIFICATION RULES BETWEEN THE MDD AND MDR

An overview of the classification rules between the MDR and MDD is presented below:

MDR		**MDD**	
Rules	**Type of Device**	**Rules**	**Type of Device**
1–4	Non-invasive devices	1–4	Non-invasive devices
5–8	Invasive devices	5–8	Invasive devices
9–13	Active devices	9–12	Active devices
14–22	Special rules	13–18	Special rules

Breakdown of each classification rule with the MDR (in comparison to the MDD).

- **Rules 1–4:** Non-invasive devices (in comparison with MDD/ AIMD).

Rule 1	**Rule 2**	**Rule 3**	**Rule 4**
No change	• Addition of "cells and tissues" to the existing language. • Blood bags moved to MDR Rule 2 from Rule 18 of MDD.	• Addition of human tissues and cells to blood, body-liquids and other liquids. • Intended for implantation or administration vs. Intended for infusion in MDD. • Inclusion of organ storage solutions, IVF media into the rule which are Class III	• Addition of injured mucous membrane to injured skin. • Replacement of 'wounds' with injuries to skin. • Also covers invasive devices that come into contact with injured mucous membrane

- **Rules 5–8:** Invasive devices (in comparison with MDD/AIMD).

Rule 5	Rule 6	Rule 7	Rule 8
• No change-clarifications only	• All devices intended specifically for direct contact with heart or central circulatory system now Class III similar to devices in contact with central nervous system	• All devices intended specifically for direct contact with heart or central circulatory system now Class III similar to devices in contact with central nervous system	• AIMD devices and accessories are Class III. • Breast implants and surgical meshes are Class III. • Total and partial joint replacements are Class III. • Spinal disc replacement implants or implantable devices that come into contact with spinal column are Class III with some exceptions (screws, wedges, plates, and instruments)

- **Rule 9–13:** Active devices (in comparison with MDD/AIMD).

Rule 9	Rule 10	Rule 11	Rule 12	Rule 13
• Addition of active devices intended to emit ionizing radiation for therapeutic purposes, including devices which control or monitor such devices, or which directly influence their performance, are classified as Class IIb. • Addition of active devices that are intended for controlling, monitoring, or directly influencing the performance of active implantable devices are classified as Class III	• Addition of 'monitoring' to diagnosis; • Active devices intended for diagnosis in clinical situations where the patient is in immediate danger as Class IIb	• New rule on software. • Classifications range from Class III–Class I	• Rule 11 in MDD. • No change	• Rule 12 in MDD. • No change

- **Rule 14–18:** Special rules.

Rule 14 (Devices with Medicinal Substances)	**Rule 15 (Contraceptive Devices, Devices for Prevention of Transmission of STDs)**	**Rule 16 (Disinfectants, Sterilizers)**	**Rule 17 (Devices for Recording X-Ray Diagnostic Images)**	**Rule 18 (Devices Utilizing Human or Animal Derivatives)**
• Rule 13 in MDD. • Clarification that medicinal product can be derived from human blood or plasma. • "Liable to act" taken out	• Rule 14 in MDD. • No change	• Rule 15 in MDD. • Addition of sterilizers to disinfectants. • Disinfectants or sterilizers become IIb only if they are used for invasive devices and as the endpoint of processing	• Rule 16 in MDD. • No change-language clarified	• Rule 17 in MDD. • Addition of cells (to tissues). • Addition of human origin cells and tissues or derivatives. • The exception about contact with intact skin only, applies only to animal tissue and does not apply to human tissues or cells

- **Rules 19–22:** Special rules.

Rule 19 (Devices Incorporating or Consisting of Nanomaterials)	Rule 20 (Body-Orifice Invasive Devices Intended to Administer Medicines by Inhalation)	Rule 21 (Devices Consisting of Substances and Introduced into the Body via Body Orifice or Skin and That are Absorbed by or Locally Dispersed)	Rule 22 (Active Therapeutic Device with an Integrated or Incorporated Diagnostic Function)
• New rule. • Classifications from III to IIa based on potential for internal exposure	• New rule. • Classification IIa or IIb. • IIb if they impact the safety and performance of the medicine or intended to treat life-threatening conditions	• New rule. • Classification from IIa to III based on where they are used and whether they or their products of metabolism are absorbed	• New rule. • Class III. • Only applies if such devices significantly determine the patient management. • Closed-loop systems or automated external defibrillators

9.4. THE GLOBAL HARMONIZED TASK FORCE

The Global Harmonized Task Force (GHTF) was founded in 1993 by various governments and medical device industry representatives from Australia, Canada, Japan, the European Union, and the United States of America. The main purpose of the GHTF is to promote and encourage an amalgamation of standards and regulatory practices that are related to the safety, performance, and quality of medical devices. For many developing or poorer countries, their policies and/or regulations governing medical devices are either inadequate or non-existent. For this main reason, the convergence of standards/regulations by the GHTF could aid those poorer countries that are importing or even manufacturing devices with:

- Information on what a major regulatory system for medical devices should look like;
- Implementing device approvals;
- Implementing health technology assessments using highly regulated markets as a guide;
- Adoption of a single medical device nomenclature;
- Advances in innovative technology;
- Implementing a proper post-market surveillance and vigilance monitoring system;
- Networking globally with regards to post market surveillance and vigilance monitoring.

The primary means by which the goals of the GHTF is accomplished is through the publication and dissemination of harmonized guidance documents that outline basic regulatory practices. The GHTF has four study groups. Group 1 is tasked with comparing operational medical device regulatory systems around the world and isolating those elements and principles that are suitable for harmonization. They are also responsible for outlining those elements and principles that may present difficulties to unifying regulations. This group is also responsible for developing a standardized format for pre-market submissions and labeling. Group 2 is tasked with reviewing current adverse-event reporting, post market surveillance and various other forms of vigilance monitoring of medical devices. This group is responsible for performing an analysis of different country requirements based on those countries with developed regulatory systems in an attempt to harmonize data collection and reporting systems. Group 3 is tasked with examining existing quality-system requirements in countries that have developed medical device regulatory systems. This group is responsible for identifying areas that are suitable for harmonization. Group 4 is tasked with examining quality system auditing practices and for developing guidance documents that combine or harmonize principles for the medical device auditing process.

9.5. INTERNATIONAL MEDICAL DEVICES REGULATORS FORUM (IMDRF)

The International Medical Device Regulators Forum (IMDRF) was conceived in February 2011 and superseded the Global Harmonized Task Force (GHTF). It was conceived as a forum to discuss future directions in the harmonization of medical device regulations. The IMDRF consists of a voluntary group of medical device regulators from around the world who have come together to build on the foundation put forth by the GHTF. The current IMDRF members include Australia, Brazil, Canada, Europe, Japan, China, Russia, and the United States of America. Official Observers include The World Health Organization (WHO) and APEC Life Sciences Innovation Forum, Regulatory Harmonization Steering Committee. In addition, the Asian Harmonization Working Party (AHWP) and the Pan American Health Organization (PAHO) are affiliate organizations which act as invited observers. Official observers do not participate in the actual decision-making process. These official observers must be approved by the unanimous consent of the Management Committee and need to be fully knowledgeable on IMDRF matters.

Regulatory convergence ideally represents a voluntary process whereby the regulatory requirements and approaches across countries and regions become more similar or aligned over time as a result of the adoption of the same technical documents, standards, and scientific principles as well as similar regulatory practices and procedures.

The IMDRF Management Committee comprises regulatory authority representatives from the following jurisdictions:

1. **Australia:** Therapeutic Goods Administration.
2. **Brazil:** National Health Surveillance Agency (ANVISA).
3. **Canada:** Health Canada.
4. **China:** National Medical Products Administration.
5. **European Union:** European Commission Directorate-General for Internal Market, Industry, Entrepreneurship, and SMEs.
6. **Japan:** Pharmaceuticals and Medical Devices Agency and the Ministry of Health, Labor, and Welfare.
7. **Russia:** Russian Ministry of Health.
8. **Singapore:** Health Sciences Authority.
9. **South Korea:** Ministry of Food and Drug Safety.
10. **United States of America:** US Food and Drug Administration (FDA).

9.5.1. Goals of the IMDRF

The main goals of IMDRF are to:

- Accelerate international Medical Devices regulatory convergence;
- Support innovation and timely access to safe and effective Medical Devices globally;
- Promote open discussion and the sharing of best practices among regulatory authorities;
- Be responsible for medical devices regulations (MDRs);
- Facilitate frequent exchange of policy and regulatory information of common interest to regulatory authorities;
- Provide opportunities to identify commonalities and develop approaches to overcome unnecessary regulatory barriers;
- Enhance communication, information sharing and scientific exchange among regulators and a broad range of stakeholders; and
- Establish development dialog with other relevant organizations.

Chapter 10

Medical Device Nomenclature

CONTENTS

10.1. NOMENCLATURE OF MEDICAL DEVICES

The nomenclature of medical devices occurs via a coding system used to generically identify such devices and health care products. The need for such a nomenclature system is essential to facilitate the management and regulation of medical devices as it allows for the standardization of various terminology used when communicating about such devices and helps to minimize any linguistic barriers that may arise. It should be noted that several naming systems exist for medical devices with a different system being used depending on the group of professionals involved (for example, procurement, stock keeping, regulatory affairs, adverse event reporting, customer services, etc.), as well as the needs of that particular group.

Having many different naming systems for medical devices can however make it difficult for interdependent communication as well as cause difficulties when individuals and/or organizations attempt to communicate. In addition, several different countries have their own nationally used nomenclature system which can further make national and international communication difficult. For this particular reason, the World Health Organization (WHO) has been working on implementing a unified nomenclature system that can, in fact, be implemented globally. The two most widely used nomenclature systems for medical devices are:

1. The global medical devices nomenclature system (GMDN); and
2. The universal medical devices nomenclature system (UMDNS).

10.2. THE GLOBAL MEDICAL DEVICES NOMENCLATURE SYSTEM

The Global Medical Devices Nomenclature System, commonly abbreviated as the GMDN, was developed by the European Committee for Standardization (CEN) together with medical device experts from around the world. This panel of experts include various manufacturers, healthcare authorities and regulators with experience in the health care industry. The GMDN is managed and maintained by a non-profit company known as the GMDN Agency, which reports to a Board of Trustees on which medical device regulators and industries are represented. In addition, to ensure long-term continuation of the GMDN, revenues are generated through the licensing and sales of GMDN Agency products, in particular GMDN codes.

Prior to the GMDN, various nomenclature systems existed, of which they were all built using different structures. Some of these nomenclature systems include the classification names for medical devices (CNMD), European Diagnostic Manufacturers Association (EDMA), ISO 9999 Technical Aids for Disabled Persons Classification, Japanese Medical Device Nomenclature (JFMDA), Norsk Klassifisering Koding and Nomenklatur (NKKN), Universal Medical Device Nomenclature System (UMDNS). Each of these systems were used either locally or nationally for a range of purposes and with varying approaches. While this was the case, all of these systems operated individually and had no impact on improving the overall communication globally or providing a common platform where medical devices could be correctly identified or safely exchanged without loss or confusion of information. The onset of the European Directives to regulate medical devices also provided an opportunity for national and international regulatory bodies to operate together and harmonize their efforts used to identify products placed on a global market.

10.2.1. Development of GMDN

Between 1993 and 1996, work done by standard organizations such as CEN and ISO led to a standard that specified an overall structure for a new nomenclature system for medical devices, which has since been revised by ISO and subsequently published as ISO 15225 Nomenclature-Medical Device Nomenclature Data Structure.

Table 10.1: Existing Nomenclature Systems Adopted and Incorporated by GMDN

Name of Nomenclature Classification System	Country
Classification names for medical devices (CNMD) and *in vitro* diagnostic products	Food and Drug Administration (FDA), USA
European Diagnostic Manufacturers Association (EDMA) *in vitro* diagnostic product classification	Used in Europe
ISO 9999 Technical Aids for Disabled Persons Classification	International use
Japanese Medical Device Nomenclature (JFMDA)	Used in Japan.
Norsk Klassifisering Koding and Nomenklatur (NKKN), Norwegian Nomenclature	Norway
Universal Medical Device Nomenclature System (UMDNS)	Developed by ECRI, USA

In 1997 a project was set up by CEN and the European Commission with the aim of creating a comprehensive nomenclature for all medical devices that could be used by interested parties globally. Six existing nomenclature systems were adopted, which allowed for coverage of a wide range of terms that defined medical devices and healthcare products. This combination of these six systems allowed for a total inclusion of 13,500 terms. These six systems which were included are defined in Table 10.1.

10.2.2. Understanding the Purpose of GMDN

The main purpose of the GMDN is to put forward a single nomenclature system that would allow regulatory authorities to regulate medical devices on a global level. In addition, the nomenclature put forward should be able to be used by healthcare providers, medical device manufacturers, importers, distributors, suppliers, and so on, as well as conformity assessment bodies and any other related parties. Thus, it can be said that the main aim of the system is to provide generic product descriptors that would ultimately be focused on allowing for global patient safety. The GMDN code is meant to allow for the international standardization of device identification together with safe data exchange between competent authorities (CA) and other parties in addition to the exchange of post-market and vigilance information, record keeping and for overall research and inventory purposes.

10.2.3. Compilation of the Initial GMDN Nomenclature Database

On 1st November 2001, the global medical device nomenclature (GMDN) was published as a CEN Report (CR 14230) and as ISO.TS 20225 with the first public release being made on a CD-ROM as GMDN version 2002.1 in November 2002 (GMDN User Guide, 2010). The initial GMDN Nomenclature database was compiled by medical device experts from around the world and was based on USO 15225. The study was directed by the European Commission so as to allow for the provision of necessary tools to allow for the carrying out of as many obligations subsequent to the implementation of the medical device directives (MDDs) in Europe. It was also to allow for the meeting of similar needs at a global level as identified in the global harmonization activities of GHTF and its various member states.

The first international workshop on harmonization of Medical Devices Nomenclature was held between the EU, EFTA, Canada, and USA in 1991 with, unfortunately, no productive outcomes. Subsequently, in 1993, CEN established a standard to define a structural basis for a device nomenclature system which allowed for the development of symbols, coding, and a proposed nomenclature. These were developed by USA workgroups in 1994 and 1995 together with a CEN recommendation for an interim nomenclature system. Following this a project proposal for a device nomenclature system was submitted in 1996. This project proposal was positively accepted by the European nomenclature standard, through the Vienna agreement; all of which resulted in a single harmonized standard. In addition, a license agreement was established between ECRI and CEN for the use of their UMDNS as a component of the development process. Subsequent to this, from 1997 to 1999, the GMDN Project began. This project was based on the developed structural standard which defined the general structure of nomenclature. The project involved approximately 70 medical device experts from 16 countries, a Project Council of 10 members, an Expert Advisory team of six members, and a secretarial team/support team of six members. This eventually led to a final GMDN Nomenclature that was put forward as a CEN Technical Report CR 14230 in 2001 and which was then publicly released. This CEN Technical Report CR 14230 was identical to the ISO Technical Specification ISO TS 20225. In addition, the GMDN Maintenance Agency Policy Group was also established. This group was responsible for future management of the GMDN in addition to developing and distributing electronic versions of the GMDN and any future versions of the data file. The British Standards Institution (BSI) has since recognized the GMDN Agency to be the formal body for the ongoing management and control of the GMDN globally.

10.2.4. Understanding the Structure of the Global Medical Device Nomenclature (GMDN) System

The GMDN system is based on and regulated by requirements stipulated in ISO 15225 which is the international standard for Medical Device Nomenclature Data Structure. This international standard specifies the rules and guidelines for a medical device nomenclature data structure to facilitate cooperation and exchange of data used by regulatory bodies on an international level. The standard contains requirements applicable to the development and maintenance of an international nomenclature system for medical device identification (Figure 10.1). GMDN data is generally defined by three levels, namely 1) device category 2) collective terms and 3) generic device group; with association with an external 4th level namely device type. Each level contains data that differs in the degree of specificity.

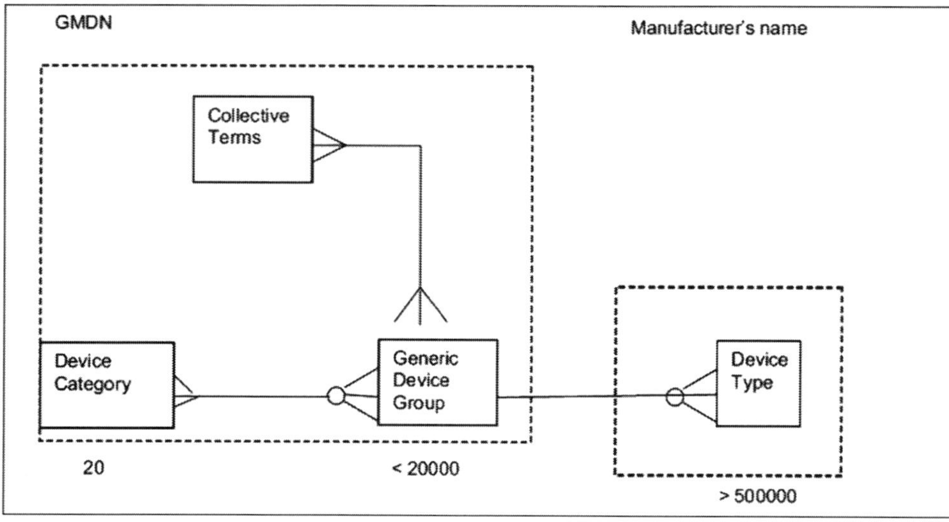

Figure 10.1: General organization of GMDN data.

10.2.4.1. Device Category

The device category forms the broadest level of the GMDN data and divides the entire medical device product market into the highest-level groups based on 1) device application, 2) technology, and/or 3) other characteristics. The standard divides or allocates the codes for 20 possible categories with 16 currently established device categories (Table 10.2).

Table 10.2: Codes for GMDN Device Categories

Code	Term
01	Active implantable devices
02	Aesthetic and respiratory devices
03	Dental devices
04	Electromagnetic medical devices
05	Hospital hardware
06	*In vitro* diagnostic devices
07	Non-active implantable devices
08	Ophthalmic and optical devices
09	Reusable devices

10	Single-use devices
11	Assistive products for persons with disability
12	Diagnostic and therapeutic radiation devices
13	Complementary therapy devices
14	Biological-derived devices
15	Healthcare facility products and adaptations
16	Laboratory equipment
17	Reserved
18	Reserved
19	Reserved
20	Reserved

10.2.4.2. Generic Device Group

The generic device group is the most significant level at which products are grouped and this occurs based on technology or intended use. Within the generic device group, there are four different types of GMDN terms which are associated with generic device groups. These groups are defined in Table 10.3 together with their alpha identifiers.

Table 10.3: GMDN Generic Device Groups

GMDN Terms	Alpha Identifiers
Preferred	P
Template	T
Synonym	S
Multiple-linked synonym	MS

In addition, there are also terms within the main database, and these are known as "equivalent terms" which form nonactive terms from the original six source nomenclatures used to develop the GMDN. These can be equal to a P, T, S, or MS term within the database and are commonly identified with the alpha identifier E. it should however be noted that a GMDN user is not to be concerned with equivalent terms as they are not active terms and have no GMDN code allocated to them. As a result, they cannot be used by a user for any practical application. E terms can however be found and viewed on the GMDN database.

10.2.4.3. Preferred Terms

Preferred terms are the only terms available for product identification and have a unique five-digit code. These terms form the optimal name selected to represent a group of medical devices or a collection of a certain type of medical devices that have the same or similar intended use or unity in technology, thereby allowing them to be grouped together without differentiation further based on individualized characteristics. Each of these preferred terms have a particular definition that describes the major characteristics of the device types in a group. These characteristics often include a physical description and an intended use or uses. In actual fact, it is the definition that determines the overall scope of the preferred term and code while the preferred term identifier is defined by (P).

10.2.4.4. Template Terms

The template term is a general device name added to the nomenclature which multiple preferred terms have identified character strings forming a base concept and which functions as a header term to create a simple hierarchy for lexically related preferred terms. Each template term has an associated definition that is inclusive of all the subordinated preferred terms. The identifier for the template terms is (T). it should also be noted that these template terms form a navigational tool only. They should not be used for identification purposes. if, however, the template terms have been used to identify products in the past, then the code has been wrongly applied and is infect, not valid. Instead, a GMDN P term and code that is applicable to the product must be selected and assigned instead. In however, no current GMDN P-code is applicable, then an application for the creation of a new term or modification of an existing P term must be submitted.

10.2.4.5. Synonym Terms

This forms the common use or familiar name used in nomenclature and is an alternative entry point in the nomenclature that is used to locate the preferred term or template term which to which it is linked. That being said, it may or may not actually be synonymous with the term to which is in fact linked. The synonym term type identifier is (S) and when selecting such a term from the GMDN database will result in automatic production of its linked preferred or template term. It must be remembered that the synonym term forms a navigational tool only and must not be used for product identification purposes.

10.2.4.6. Multiple-Linked Synonym Terms

The multiple-linked synonym term forms a medical device name, and this arises often from one of the GMDN source nomenclatures. It is usually of a higher order and is often linked to more than one preferred term and/or template term. It should be noted that the multiple-linked synonym term forms a navigational tool only and must not be used for product identification purposes.

10.2.4.7. Collective Terms

Collective terms are used to group those medical device terms that have common features, and which can be applied to a whole range of subject matters. Each collective term is assigned a four-digit incremental code with the prefix (CT). Subject matters applied include for example:

- To illustrate the scope of certificates issued by notified bodies (NB). This is applicable when assessing which groups, families or types of medical devices are covered within a manufacturer's quality system.
- To identify the range of skills and general technological abilities for which a notified body has been approved and appointed by the relevant competent authority.
- For exchanges of information between CA when general information on individual manufacturers capabilities is notified for inclusion. in the European database for medical devices (EUDAMED).

Several means have been identified to group GMDN generic medical device group terms using collective terms. These include:

- Those medical devices covered by the application of common technology;
- Those medical devices manufactured using similar manufacturing procedures and common technical features;
- Those medical devices manufactured for the application of similar medical procedures;
- Those medical devices manufactured using common materials requiring special skills;
- Those medical devices developed to meet specific risk-associated considerations.

10.2.5. Understanding the Coding System Used for GMDN

All terms within the GMDN are assigned a unique code which provides a form of security to avoid misunderstandings, language barriers, or any other kind of discrepancies within the database. This unique code consists of five digits starting from 10000. These codes can be divided into one of three categories that make up the coding system for the GMDN. These include the following subsections.

10.2.5.1. Codes in the Range of 1–9999

These codes are not represented in the GMDN and are exclusively reserved for assignment by any end-user. It may be used as desired in any user's local data system. These codes should not be used for any kind of official purpose as this could lead to ambiguity.

10.2.5.2. Codes in the Range of 10000–30000

These codes are represented in the GMDN. They have been reserved to represent the original code used by the ECRI organization for their UMDNS terms that have since been adopted for use in the GMDN. In essence, this will allow a GMDN user to access an automatic mapping-like system which links the ECRI code representing the UMDNS term to the identical GMDN code used currently to represent the GMDN term. Ideally this will assist in the transition for users who have previously used the ECRI UMDNS terms.

10.2.5.3. Codes above the Range of 30000

All of these codes for GMDN codes which have been created for the GMDN database.

10.2.6. Understanding the Coding System Used by the GMDN

The coding system used by the GMDN is explained in detail in Box 10.1.

Box 10.1: Understanding the GMDN Term Structures

Each GMDN term consists of three parts:

- **Term Name:** General-purpose syringe.
- **Definition:** A sterile device that consists of a calibrated hollow barrel (cylinder) and a moveable plunger intended to be used to inject fluids (e.g., medication) into, and/or withdraw fluids/gas from, the body or a medical device for various medical.
- **Code:** 47017.

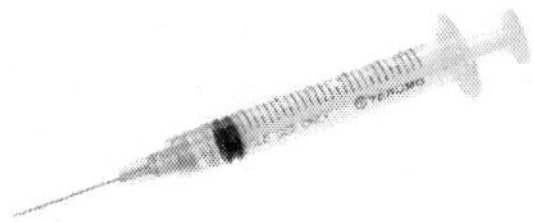

10.2.7. Services Provided by the GMDN Agency

The GMDN agency provides numerous services such as:

- Access to the GMDN data file and codes using an Internet site, a license agreement as well as direct credit card purchase.
- A link from the GMDN database to the user's in-house data system through a license agreement and a custom-made software link.
- Application form for new terms or modifications to existing terms and/or definitions used in the identification of a product.
- Access to the GMDN terminology and information.
- Guidance on how to use the GMDN Database.
- Access to collective terms.
- GMDN translation software tools.

The main role of the GMDN Agency is:

- To carry out the functions of a Maintenance Agency to develop and maintain the "GMDN" for medical devices.
- To meet the needs identified by the European Commission when developing the European Directives on Medical Devices such that it allows for the establishment of a nomenclature for medical device generic descriptors that will meet a global need around suitable identification purposes.

- To ensure that the GMDN meets the needs of all national authorities, industries, and other global users.
- To ensure that the GMDN is constructed with reference to European and International Standards.
- To be responsible for adding, amending, and archiving terms and definitions for medical devices and to assign codes as and when required, thereby allowing for easy identification. To also allow for consistency such that all translated versions of the nomenclature will carry an identical code for each generic term.
- To liaise with standard bodies and be aware of the current standards on medical device nomenclature.
- To coordinate and link with appropriate organizations for the translation of GMDN into other organizations.

10.2.8. Benefits of the Global Medical Device Nomenclature (GMDN) System

Have a global system for medical device nomenclature allows for common terminology to be used between global regulatory partners, thus allowing for efficient communication and adequate and correct sharing of medical device details. Nationally, it allows for the use of common terminology, which in turn allows for regulatory bodies as well as individual industries such as manufactures, distributors, and importers to efficiently communicate and share accurate information. The main aim of the GMDN agency is to provide guidelines for the appropriate and consistent name of various medical device groups as well as the creation of a descriptive set of definitions that are device-group specific. In this way, the use of data arising from the GMDN agency will help regulators to improve their nomenclature practices as well as enforce consistency across generations and ensure proper data management, which could influence other areas like post market activities and adverse event reporting.

10.2.9. How to Use GMDN Codes?

Use of the GMDN codes begins by initially identifying the codes for the product in question from the GMDN database. This code is provided to relevant customers and used to register the product with a unique device identifier (UDI) (Box 10.2).

Box 10.2: Using the GMDN Codes Explained

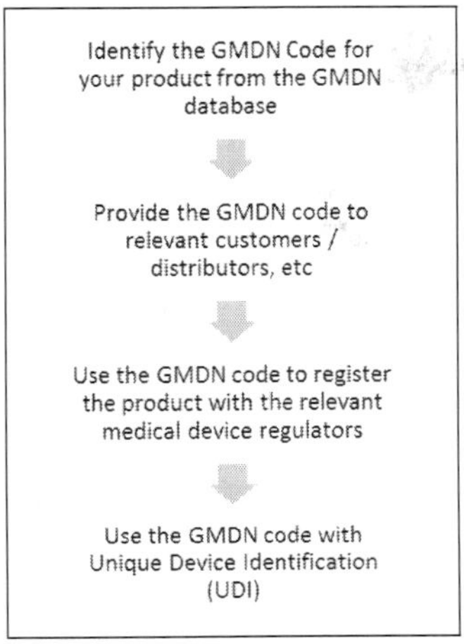

10.2.10. Application of the GMDN to International Use

As health care treatment continues to grow and globalize, so does the universal need for patient safety. This has subsequently created a need for addressing, tracking, and tracing medical devices at a global level. The GMDN is used in the European Databank on Medical Devices, better known as EUDAMED which was established by the European Commission. The main purpose of EUDAMED is to strengthen the global market surveillance and vigilance monitoring of medical devices. The UDI system is of the ways that has been developed to trace and track medical devices, with the main aim being to increase patient safety while also working to improve overall market surveillance. The aim is for this to be done by ensuring a globally harmonized and consistent approach through implementing harmonized legislation for UDI using global standards.

10.3. THE UNIVERSAL MEDICAL DEVICES NOMENCLATURE SYSTEM (UMDNS)

The UMDNS was initially developed by the Emergency Care Research Institute (ECRI), which is a non-governmental, non-profit organization, governed by an Executive Committee and a Board of Trustees. The UMDNS was developed as a standard international nomenclature and computer coding system for medical devices with the main purpose of facilitating the identification, processing, filing, storage, retrieval, transferring, and communication of data pertaining to medical devices. The UMNDS contains over 12,000 preferred terms and over 28,000 synonyms, entry terms and related terms. All terms are assigned a 5-digit code using consecutive numbering with no intrinsic meaning. The code itself is associated with a definition and a description of the intended use. The database is maintained by a core group of ECRI nomenclature specialists and the database itself is updated monthly. The nomenclature is used in a variety of applications ranging from hospital inventory to work-order controls to national regulatory systems to procurement to medical device databases.

10.4. SAFETY OF MEDICAL DEVICES

All medical devices, regardless of the class, possess a certain degree of risk associated with it, and while this is the case, many of the problems cannot be detected until in-depth market-based experience is gained. More simply put, the safety and risk of the device is judged when that device fails to respond in a manner that it should have, or rather when a device functions in a manner that was not predictable at the time of use or due to unique conditions in the patient itself. For many manufacturers, the current approach is to carry out a suitable risk assessment which involves determination or, rather, estimation of the potential of a device becoming a hazard that could subsequently lead to additional problems or harm to end-users.

10.5. SAFETY IN RELATION TO PERFORMANCE OF A MEDICAL DEVICE

Clinical effectiveness is a fairly good indicator of how a device performs, thus if a device is said to be clinically effective, it means it produces the effect intended by the manufacturer relative to the medical condition the manufacture claims it will treat. It is much easier and more accurate to measure performance rather than clinical effectiveness. Performance may include technical functions and is very closely linked to the safety of a device, as any medical device that does not perform well could lead to serious clinical safety issues for a patient. For this reason, safety, and performance of medical devices tend to be considered closely together.

10.6. A SIMPLE MEDICAL DEVICE INFOGRAPHIC ON HOW TO CLASSIFY MEDICAL DEVICES?

With the theory behind how to classify medical devices, it is easy to get confused, so looking at a summary infographic may help. Below is a simple infographic designed by www.scientistssanctuary.com to help organizations classify their medical device based on risk. It also provides an overview of the assessment route a manufacturer will have to take (Figure 10.2).

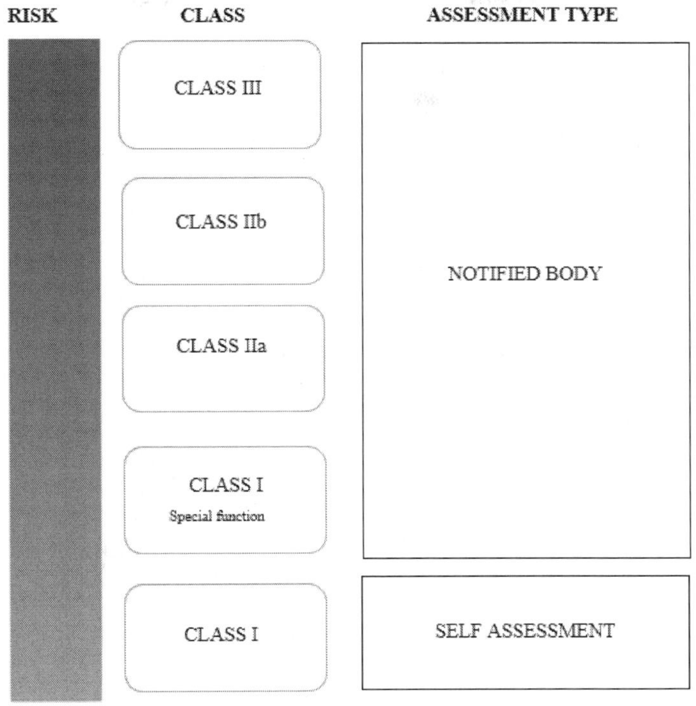

Figure 10.2: Simple infographic on how to classify medical devices.

Source: Scientists Sanctuary have also designed an easy-to-use checklist to help medical device organizations determine their device classification. You can find this checklist at www.scientistssanctuary.com.

10.6.1. A Simple Method to Classify Medical Devices

A simplified three-step method to classify your medical device is presented below (Figure 10.3):

1. Decide which category the device belongs to by looking at all the definitions within the regulation;
2. Look at each rule within the regulation. If many rules are applicable, you should choose the one with the highest risk; and

3. If your product is a borderline product (i.e.: a product that is a combination of a drug-device), you need to identify the main intended use.

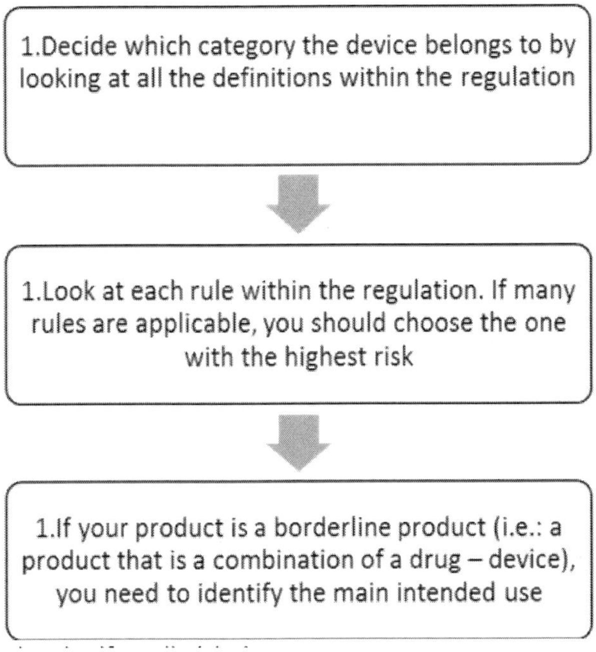

Figure 10.3: Simple method to classify medical devices.

10.7. BORDERLINE MEDICAL DEVICES

Understanding the difference between a Medicinal Product and a Medical Device (MD) is imperative to launching a new product on the market. Understandably, it is also one of the areas of biggest concerns and confusion for manufacturers, importers, and distributors.

Article 2(2) of the European Directive 2001/83/EC relating to medicinal products for human use indicates that respective Medicinal Products Legislation must be applied whenever a product can be covered by both the definitions for medicinal products and for products regulated by other legal provisions enacted by the European Community.

Article 2: DIRECTIVE 2001/83/EC on the Community Code Relating to Medicinal Products for Human Use
1. This Directive shall apply to medicinal products for human use intended to be placed on the market in Member States and either prepared industrially or manufactured by a method involving an industrial process. 2. In cases of doubt, where, taking into account all its characteristics, a product may fall within the definition of a 'medicinal product' and within the definition of a product covered by other Community legislation, the provisions of this Directive shall apply.

This article implies that the decision to base the risk-benefit assessment of the product in question on the MDD would contradict the aforementioned constitutional principle, pursuant to which the stricter of the regulatory procedures theoretically possible is to apply in cases of doubt. It should also be noted that in contrast to the approval procedure established for medicines, the MDD requires a conformity assessment procedure to be performed by the manufacturer and a notified body. To avoid such discrepancies in overall procedures, it is imperative that the classification of the device be carried out in full compliance with existing regulations that also take into account related guidance documents that have been issued by the European Commission.

10.8. DISTINCTION BETWEEN A MEDICINAL PRODUCT AND A MEDICAL DEVICE

The European Directive 2004/27/EC which is an amendment of Directive 2001/83/EEC, defines medicinal products as:

- "any substance or combination of substances presented as having properties for treating or preventing disease in human beings; or
- any substance or combination of substances which may be used in or administered to human beings either with a view to restoring, correcting, or modifying physiological functions by exerting a pharmacological, immunological or metabolic action, or to making a medical diagnosis."

Medical devices, on the other hand, have been previously defined in this volume and can be grouped into Class I, II, and III devices.

- Class I devices are usually subjected to limited regulatory supervision due to the low risk of the product and simpler intended use. A detailed conformity assessment procedure should be carried out, however. Examples of Class I devices include elastic bandages and examination gloves.

- Class IIa devices (non-invasive products or invasive products for short-term use) and Class IIb devices (invasive products for long-term use, including use on wounds) are those where general supervision alone is enough to ensure safety and efficacy. For this reason, systems need to be in place to ensure safety and efficacy are achieved. Devices intended specifically to be used for disinfecting medical devices are grouped as Class IIa. All devices intended specifically to be used for disinfecting, cleaning, rinsing, or hydrating contact lenses are in Class IIb. Other examples for Class IIb products are blood bags and devices that are intended for use principally on wounds that have breached the dermis and can only heal by secondary intent. Devices are subject to categorization into Class IIa if they are intended to manage the microenvironment of a wound.
- Class III is the most stringent regulatory category for medical devices. Class III devices are usually products that either support or sustain human life, are of substantial importance in preventing impairment of human health, or the use of which, present a particular potential risk of illness or injury. Examples of Class III devices are all implantable devices for direct contact with the heart, the central circulatory system, or the central nervous system, as well as other implantable devices and long-term surgical devices to administer medical products.

It should be noted that the intended purpose, which is defined as "prevention of disease" applies to both medical products and devices. A common example where confusion can arise is with hand disinfectants. Currently, hand disinfectants are being marketed in Germany as medical products, in other European countries like Austria, Bulgaria, Lithuania, and Poland as Biocidal Products and in some countries as medical devices.

The European Commission has released and since updated their "Manual on Borderline and Classification in the Community Regulatory Framework for Medical Devices" which is intended to assist manufacturers in determining whether their product falls within the definition of a medical device laid down in the Council Directive 93/42/EEC for medical devices.

In general, a product will be considered to fall within the definition of a medical device if it has a medical purpose and if the product functions primarily in a way that is not metabolic, immunological, or pharmacological. Determination of whether a product is considered to have a medical purpose will then be based on its intended purpose. The MDD provides for several rules for the exact classification of a medical device. The Borderline Manual provides guidance concerning a broad range of "borderline" products such as water filters, shoe covers, radiation shields, fluid collection bowls and hand disinfectants. The European Commission has updated the Borderline Manual with guidance concerning the classification of three products. The three products are automated external defibrillator storage units, lubricants for the alleviation of vaginal dryness and medication decisions support software.

More detail on the manual is provided below:

The manual defines borderline cases as being those cases where it is not clear from the outset whether a given product is a medical device, an in vitro diagnostic (IVD) medical device, or an active implantable medical device or alternatively, those cases where the product falls within the definition of a medical device but is excluded from the Directives by their scope. In such cases where a given product does not fall within the definition of medical device or is excluded by the scope of the Directives, other Community and/or national legislation may be applicable. There may be cases where 'claims' of a medical nature are made for certain products and where those claims cannot be substantiated by technical, clinical, and scientific data. If there is insufficient clinical, technical, and scientific data to support the claims made, the product would not meet the requirements of the MDDs and therefore may not be CE marked as a medical device. For such products no medical claim can be made.

The manual further provides details on various borderline devices which are discussed below. Background information on how the device is used is provided. It should be remembered that defining a given product as a medical device and interpretation of the application of the classification rules falls within the competence of the CA of the Member States where the product is on the market. Different interpretations of legislation may occur, and for this reason, the European Commission has highlighted the importance of facilitating dialog among regulators and industry.

10.9. MEDICAL DEVICE/*IN VITRO* DIAGNOSTIC (IVD) MEDICAL DEVICE-INTENDED MEDICAL PURPOSE

Let us recap the definitions according to the Regulations:

According to Article 1(2)a MDD "'*medical device' means any instrument, apparatus, appliance, material, or other article, whether used alone or in combination, including the software necessary for its proper application intended by the manufacturer to be used for human beings for the purpose of:*

- *diagnosis, prevention, monitoring, treatment, or alleviation of disease;*
- *diagnosis, monitoring, treatment, alleviation of or compensation for an injury or handicap;*
- *investigation, replacement, or modification of the anatomy or of a physiological process;*
- *control of conception, and which does not achieve its principal intended action in or on the human body by pharmacological, immunological or metabolic means, but which may be assisted in its function by such means.*"

According to Article 1(2)b IVDD "'*in vitro diagnostic medical device' means any medical device which is a reagent, reagent product, calibrator, control material, kit, instrument, apparatus, equipment, or system, whether used alone or in combination, intended by the manufacturer to be used in vitro for the examination of specimens, including blood and tissue donations, derived from the human body, solely or principally for the purpose of providing information:*

- *concerning a physiological or pathological state, or*
- *concerning a congenital abnormality, or*
- *to determine the safety and compatibility with potential recipients, or*
- *to monitor therapeutic measures.*"

From this definition, it follows that in order to fall within the definition of an IVD medical device, the product must also meet the definition of a medical device.

Examples as given by the manual are discussed below:

Borderline Device	Background	Outcome
Lightbox indicated to treat seasonal affective disorder (SAD)	The product in question is a lightbox that emits bright light, and the manufacturer states that light therapy is 'a convenient and effective way of compensating for the lack of light without resorting to medication.' The manufacturer also states: 'in autumn and winter, the seasons with the least sunlight because the days are shorter, increased symptoms resulting from light deprivation may be experienced. Even standard artificial lighting in buildings cannot compensate for a shortage of natural light. The consequences of this may be depression, lack of drive, interrupted sleep and melancholia-the typical autumn-winter blues.'	These statements are effectively claiming for treatment of seasonal affective disorder (S.A.D.), which is a generally recognized medical condition, and therefore this product is considered a medical device.
AB0 and Rhesus (D) blood grouping intended for diet purposes	These products are tests for AB0 and Rhesus (D) Blood Grouping, which is sold through the internet and which is used by laypersons in the home environment. The manufacturer states the following: HOME-KIT, for *in vitro* diagnostic use, not for bed-side testing; and currently in the text, which describes interpretation of the result, the manufacturer states not for clinical use. The manufacturer has stated that the main reason and purpose of the tests is educational. It is indicated that the product enables the user to ascertain their blood group in order to determine whether a specific (food) diet should be followed. This decision was not related to following a specific diet for medical purposes.	According to the information given by the manufacturer, it is concluded that even though it can be argued that this product fits some parts of the definition of an *in vitro* diagnostic medical device, it does not meet the definition of a medical device. As the definition on *in vitro* diagnostic medical device (Article 1(2) b IVDD) reads "'*in vitro* diagnostic medical device' means any medical device, the product must also meet the definition of medical device. It can be concluded that as the intended purpose of this product cannot be qualified as a medical purpose as described in the definition of a medical device (Article 1(2) an MDD), this product is not an *in vitro* diagnostic medical device. The product in question is not a blood typing test for a medical purpose. This conclusion is reached in the light of the information provided by the manufacturer. It would be necessary to check if these statements are correct, and, consequently, do not contain a deceptive and misleading labeling. Therefore, no reference to *in vitro* diagnostic medical device (e.g., 'for *in* vitro diagnostic use') can be made. Also, as the product is intended to be used by laypersons, there is a need for a strong and clear disclaimer which is understandable for lay users; i.e., a statement that the test results cannot be used for transfusion purposes or for blood group determination for medical purposes. It has to be noted that only a statement that the product is not a medical device can not constitute a reason to escape from the Directive and to avoid the CE marking if the criteria of the definition of a medical device are satisfied.

Pharmacy compounders	Pharmacy compounders may be used in a hospital pharmacy or in an industrial environment. They are products that are intended for the production of fluids for administration to a patient, usually as intravenous fluids (IV fluids) for administration parentally. They are intended to be used by clinical nutritionist specialists and pharmacists. They mix a number of ingredients that are subsequently administered to the patient. These fluids may be nutritional solutions or pharmaceuticals. They may also be used for compounding formulas for cardioplegia, hydration, fluid drugs and renal replacement therapy. The issue is whether pharmacy compounders are medical devices, particularly if the compounder is specifically intended to be used by clinical nutritionist specialists and pharmacists to correctly prescribe nutritional solutions for Total Parenteral Nutrition.	Pharmacy compounders are regarded as processing equipment and should not be qualified as medical devices. The compounder is only used for mixing the solution to be administered to patients, but it does not administer itself anything to a patient. It should therefore not be qualified as a medical device.
Dental disclosing products	Dental disclosing products are intended to 'disclose' plaque, i.e., to highlight the areas around the teeth where the plaque is in order to aid its removal. There may be claims to 'aid oral hygiene,' to 'aid correct brushing regimes' or simply to identify the plaque for its removal. Dental disclosing products may be in the form of solutions, tablets or an applicator containing the solution and may be intended for use by dentists or by individuals at home. The question is to whether these products should be qualified as medical devices, or whether they are simply intended for oral hygiene and therefore shall not be considered as medical devices.	Although in severe cases, in addition to other contributory factors, plaque may lead to dental decay or gum disease, plaque is not considered to be a disease in its own right. Therefore, dental disclosing products, intended to disclose plaque in order to help its removal, cannot be qualified as medical devices.

Mixer	The product is a Thermomixer (mixer) intended to control the temperature of (and mix) liquids in closed micro test tubes and micro test plates. The manufacturer claims that this mixer is specifically intended for the preparation and processing of samples from the human body within the scope of *in-vitro* diagnostic applications, in order to allow the *in-vitro* diagnostic medical device to be used as intended. Therefore, the manufacturer considers this Thermomixer, being an *in-vi*tro diagnostic accessory, to fall within the scope of Directive 98/79/EC.	MEDDEV 2.14/1 rev.1, states that "if, however, the product does not in fact possess specific characteristics that make it suitable for one or more identified *in vitro* diagnostic examination procedures, then the manufacturer is not free to bring it within the scope of the IVDD merely by affixing the CE marking to it. In other words, a manufacturer is not able to bring within the scope of the IVDD a product that, in reality, is a piece of general laboratory equipment simply by affixing the CE mark to it." For example, the point 4 of MEDDEV 2.14/rev.1 mentions that laboratory centrifuges are not usually considered to fall within the scope of the IVD directive. According MEDDEV 2.14/1 rev.1, the fact that this mixer is intended by the manufacturer to be used especially for *in vitro* diagnostic procedures is not sufficient to qualify it as an IVD medical device, if this mixer does not possess specific characteristics that make it suitable for one or more identified *in vitro* examination procedures. If this mixer possesses such specific characteristics, the manufacturer will have to demonstrate these specific characteristics and the link with one or more identified *in vitro* examination procedures. This case is similar to the cases of pipettes or glass slides, considered as products for general laboratory use, already published in the manual. On the basis of the above, this mixer could not be considered as an *in vit*ro diagnostic medical device.
Non-corrective contact lenses	In general, non-corrective contact lenses (commonly known as 'plano' lenses) are not considered to be medical devices as they have no corrective function. Some non-corrective contact lenses, colored or not, however may have a medical purpose. They serve to treat a number of congenital or traumatic conditions. They are often used in clinical practice or post-surgical setting as a medical prosthesis. Examples of non-corrective contact lenses with medical purpose are: • UV blocking contact lenses that help protect against transmission of harmful UV radiation to the cornea and into the eye so as to alleviate photophobia as seen in albinism. • Contact lenses for therapeutic use as a bandage lens for the following acute and chronic ocular conditions:	All of the above are recognized medical conditions and therefore non-corrective lenses, colored or not, which have the above-mentioned functions are considered to be medical devices on the basis that they prevent, monitor, treat or alleviate disease. Manufacturers of such products must, however, clearly indicate the specific intended medical purposes for these non-corrective contact lenses both on the packaging/labeling and in the instructions for use. Non-corrective contact lenses without specific claims (such as those listed above) would be regarded as having no medical purpose and therefore not medical devices.

	• For corneal protection in lid and corneal abnormalities such as entropion, trichiasis, tarsal scars and recurrent corneal erosion. In addition, they are indicated for protection where sutures or ocular structure malformation, degeneration or paralysis may result in the need to protect the cornea from exposure or repeated irritation; • For corneal pain relief in conditions such as bullous keratopathy, epithelial erosion and abrasion, filamentary keratitis, and post keratoplasty; • For use as a barrier during the healing process of epithelial defects such as chronic epithelial defects, corneal ulcer, neurotrophic, and neuroparalytic keratitis, and chemical burns; • For post-surgical conditions where bandage lens use is indicated such as post-refractive surgery, lamellar grafts, corneal flaps, and additional ocular surgical conditions; • For structural stability and protection in piggy back lens fitting where the cornea and associated surfaces are too irregular to allow for corneal rigid gas permeable (RGP) lenses to be fitted. In addition, the use of the lens can prevent irritation and abrasions in conditions where there are elevation differences in the host/graph junction or scar tissue.	
Biofunctional clothes	Biofunctional clothes or "therapeutic clothes" consist of clothes (e.g., socks, leggings, pajamas, undershirts…) impregnated with silver and brown algae extract. These clothes are presented as having anti-bacterial, anti-microbial, breathable, thermal regulating and anti-odor properties. According to the manufacturer, the silver ions (positively charged) integrated into the fiber reduce the microbial growth in the fabric and inhibit microbial growth on the skin, and the brown algae extract protects against rash act by neutralization of neuromediators and vasodilator agents responsible for flushing. The intended use of these clothes is to prevent inflammatory crisis of atopic dermatitis.	The clothing in itself cannot be considered as a medical device unless it achieves a specific medical purpose. Silver is a medicinal substance, presented as having anti-microbial properties. The combination of clothes with a medicinal substance might not be qualified as a medical device. In addition, the intended use described by the manufacturer, prevention of inflammatory crisis of atopic dermatitis, is likely to be achieved by pharmacological or immunologic means and linked to the presence of silver and algae extract. Therefore, the clothes would not be considered as a medical device, based on the determination of the principal mode of action. In case the clothes would have in themselves a medical purpose and the principal mode of action would not be achieved by metabolic, pharmacologic, or immunologic means, the product might be seen as a medical device. However, in that case, the medical device would fall in Class III under Rule 13, due to the combination with a medicinal substance.

	The main mode of action described by the manufacturer is the creation of a physical barrier that prevents the contact of sensitive skin with the outside and other tissues potentially sensitizing, creating an environment helping to the attenuation of skin conditions. The actions of silver and algae extract were presented as an ancillary action. The product was presented as a Class I medical device.	
System for the determination of bacterial contamination in blood products	The bacterial contamination of blood components represents a high infectious risk in blood transfusion, particularly for platelets concentrate. In order to estimate the bacterial contamination, manufacturers have developed some bacterial detection systems near the time of blood collection. A new system has been developed based on the detection and identification of a wide spectrum of common pathogenic bacteria. Following the identification, a direct count is obtained by cytometry.	These devices are intended to reduce the risk of transfusion reaction due to the bacterial contamination. The system is providing information regarding the safety of blood donation and therefore should be qualified as an *in vitro* diagnostic medical device.
Independent *in vivo* dosimeters	*In vivo* dosimetry systems are used in radiotherapy to monitor the radiation dose received by the patient during a radiotherapy treatment. The device consists in a semi-conductor detector placed on the patient's skin and an electrometer placed in the control room where the physicians control the radiotherapy equipment. When performed early in treatment as an additional measure, the *in vivo* measurements of the radiation dose are an additional safeguard against setup, calculation, or transcription errors. If the dose delivered to the patient is higher than the calculated dose, the system, independently of the radiotherapy equipment, can alert the physician, allowing him to act on the treatment.	Even if the *in-vivo* dosimeter is independent of the radiotherapy equipment, it can be considered that it is used to control the performance of the radiotherapy device. Therefore, the *in-vivo* dosimeter should be qualified as a medical device. The *in vivo* dosimeter controls the performance of an active device intended to emit ionizing radiation and therefore should be classified as Class IIb, according to Rule 10.

Gallipots	Gallipots are containers usually made from metal or plastic. They may be sterile or non-sterile and may be disposable. They are intended to be used to contain various items, including medicinal products or disinfectants for antiseptic fluids to scrub the incision area prior to surgery. They are defined variously as "a small glazed pot used by apothecaries for medicines, confections, or the like," as "a small glazed earthenware jar formerly used by druggists for medicaments," and as "a small usually ceramic vessel with a small mouth; especially: one used by apothecaries to hold medicines."	Gallipots do not meet the definition of a medical device: They do not diagnose, prevent, monitor, treat or alleviate disease, diagnose, monitor, treat or alleviate or compensate for injury or handicap or investigate, replace, or modify the anatomy or physiological process. Nor do they fit the definition of an accessory to a medical device. Therefore, Gallipots are not considered to be medical devices or accessories to medical devices.
Shoe covers	On the EU market there are shoe covers sold at hospitals used for different purposes: • Shoe covers which are intended by their manufacturer to be worn by the nursing staff to limit cross-contamination in operating theater and care units. • Shoe covers for visitors which are intended to be worn by visitors to prevent patient's contamination when they visit a patient.	According to the Section 3.3 of Meddev 2.1/4, the labeling of the product is crucial for its classification under the directive 93/42/EEC. "As a general rule, the principal intended purpose can be established as being the one of a medical device if the product is intended to be used in a medical context with the aim to provide protection of health and safety for the patient, regardless of whether the product aims simultaneously to protect also the user. Where a product is mainly intended to protect the person using it, irrespectively whether in a medical environment or not, it falls under Directive 89/686/EEC." Consequently, shoe covers which are specifically intended by their manufacturer to be used in operating rooms, intensive care units, or immunodepressed patients to protect the patient from potential contamination are medical devices. In application of Rule 1, they are in Class I. In this case, shoe covers are considered as being similar to surgeons' gowns and hats. On the other hand, shoe covers for visitors even in a hospital are products of control of the environment.
Urine Diverter/ Funnel Element for Mid-Stream Urine Collection	There are a number of midstream urine collection kits on the EU market. In general, these products consist of a specimen receptacle and a urine diverter/funnel element. They are usually packaged together in a 'kit' and in general the funnel part is not attached to the specimen receptacle in the kit. The user puts the two parts together, produces the sample, then removes the 'funnel' and closes the specimen receptacle. The specimen receptacle is an IVD medical device as it is intended for the primary containment and transport of the urine sample.	The funnel/diverter itself does not fit the definition of an IVD medical device: it does not directly diagnose or provide information. It has no diagnostic function in itself. It therefore cannot be an IVD medical device in its own right, but is regarded as an accessory to the specimen receptacle. Directive 98/79/EC states in Article 1.2 (c) that for the purposes of the definition of an accessory, invasive devices or those which are 'directly applied to the human body' are not considered to be accessories to IVD devices, therefore the 'funnel'/diverter cannot be considered to be an accessory to the specimen receptacle. MEDDEV 2.14/1 states that specimen receptacles which come into contact with the patient fall within the MDD and not the IVDD. It also reiterates Article 1.2 (c) of the Directive and states that devices which

	The question arose as to the qualification of the urine diverter or 'funnel' and whether it is a general medical device (Class I) or an IVD medical device. Some of these 'funnels' are intended to be held away from the body, whilst others are intended to be applied to the body. Some of these have been specifically designed for female use, and this can be seen in their shape and form. In the majority of these cases, the intention is for the 'funnel' to be held against the body whilst the urine specimen is collected.	are made available with IVD devices and which are directly applied to the human body for the purpose of obtaining a specimen within the meaning of Directive 93/42/EEC are not considered to be accessories to IVDs and come within the scope of Directive 93/42/EEC. Therefore, where the funnel/urine diverter is intended to be directly applied to the human body for the purpose of obtaining a specimen (within the meaning of Directive 93/42/EEC) it shall not be an accessory to an *in vitro* diagnostic medical device but it will be a Class I medical device. The intended direct application to the female/male body for obtaining the specimen is demonstrated by the instructions for use, the labeling and/or other information provided by the manufacturer.
Air purifiers/ Air decontamination units/ Mobile air decontamination units	There are various types of air purifiers and air decontamination products on the market. There are two main types: • 'Standalone'/mobile units which are intended to purify or decontaminate the air in individual rooms or areas, which may be moved from room to room in accordance with the perceived need to purify or decontaminate the air in the room. They stand in the room and function independently. They are not connected to individual patients directly (e.g., via the use of a mask). These may or may not include filters. • Units which are installed into the fabric of the healthcare institution, supplying purified/decontaminated air through pipe works that are part of the fabric of the building, into specific rooms/areas. Filters may also be used within the systems. The claims for these products vary from 'purifying' the air to 'decontaminating' the air; however, in general, they are intended to remove allergens (dust, pollen, etc.), and/or to remove bacteria from the air in a specific room or area. Some of these systems contain filters in order to remove particulates in the air, whilst others claim to destroy airborne micro-organisms. In both cases, the purified air is supplied into rooms and not directly connected to individual patients (e.g., via a mask) nor to breathing machines (which are directly connected to a patient).	These products are intended to 'control the environment' by removing allergens or microbial contamination from the air. They do not act directly on an individual patient, and there is no direct contact with an individual patient. In order for a product to be a medical device, the device must have a direct association with the individual patient. Although maintaining clean air may contribute to keeping a patient in an appropriate environment, this is not considered to be a 'medical purpose.' Air is part of the environment of the patient, and its cleanness is necessary in a similar way as for surfaces, walls, floors, and other objects which also need to be cleaned and disinfected. Since these products do not fulfill the definition of a medical device, they are not considered to be medical devices, but are rather products for the general environment.

	In hospitals, these systems for air decontamination are used to reduce infection risks in intensive care units and operating theaters. They include Burnt units or systems comprising a mobile air decontamination unit and a mobile deployable room put directly above the patient with the intent to protect burnt patients or immuno-compromised patients from air contamination. The question has arisen as to whether such products are considered to be medical devices since it is claimed that they reduce infection risks to patients or protect patients from airborne contami-nants.	
Wigs and headscarves	Both wigs and headscarves are rou-tinely used by people for esthetic reasons. They may also be used by patients undergoing chemotherapy or suffering from alopecia (hair loss) from other causes. The question arose whether wigs and headscarves intended for such patients would be regarded as medical devices on the basis that they may protect the head, reduce physical or emotional effects of hair loss, and so on.	Wigs and headscarves are primarily intended for a cosmetic purpose, i.e., to improve or change the ap-pearance of the wearer. They do not treat or alleviate any specific medical conditions and do not fit the definition of a medical device. They are therefore not qualified as medical devices.
Blood irradia-tion indicators	Lymphocytes contained in the blood components can bring to Transfusion Associated Graft-versus-Host Disease (TA-GvHD) in specific cases. Lymphocytes may be deprived of the possibility of multiplication through the use of irradiation. Therefore, for example, red cells for intrauterine transfu-sion and whole blood for exchange transfusion in neonates should be irradiated prior to transfusion. Blood irradiation indicators provide information as to whether the blood products have been ir-radiated and protect from repeated irradiation of those products. According to instruction for use, indicators only indicate that ir-radiation has occurred, while they do not measure the dose from an irradiator. The manufacturer of the above-mentioned indicators has placed these products on the market as Class I medical devices.	The blood irradiation indicators simply show that blood product has been irradiated. They do not affect the irradiation process and only provide additional information to the user. In this connection, blood irradiation indicators do not fulfill the definition of a medical device laid down in Article 1(2)(a) of Directive 93/42/EEC; therefore, they are not considered to be medical devices. Blood irradiation indicators cannot be considered to be accessory to a medical device because they are not intended specifically by its manufacturer to be used together with a device to enable it to be used in accordance with the use of the device intended by the manufacturer of the device.

Odor neutralizers	Odor neutralizers, which may be presented in a spray form or as oil, are products used for the control of odors generated by the ostomy devices. The spray is intended to neutralize odors in room wherein use ostomy equipment is present, while the oil is intended to be instilled in ostomy pouches. These products are intended to improve the quality of life of people holding ostomy pouches by controlling the odors.	It is considered that these products have no medical purpose and therefore cannot be qualified as medical devices. The sprays are not considered to be medical devices-they are intended to be used in a room and do not fit the definition of a medical device. The oils meet the definition of an accessory to a medical device only when these articles are intended specifically by their manufacturer to enable the pouches to be used in accordance with their intended purpose, as specified by their manufacturer, according to the definition of accessory in Article 1 par. 2(b) of Directive 93/42/EEC.
Bedwetting alarm	A bedwetting alarm is a device intended to be used as a treatment of nocturnal enuresis. Each device has a moisture sensor component and an alarm component. A moisture sensor snaps onto underwear. The alarm sounds at the first sign of moisture, and interrupts the wetting process. The bedwetting alarm conditions the user to timely recognize bladder fullness, and awaken before wetting occurs.	When such devices are intended for the treatment of nocturnal enuresis, which is generally recognized as a medical condition, they should be qualified as a medical device. Such bedwetting alarm devices should be classified as Class I under Rule 12 of Annex IX of the Directive 93/42/EEC.
Sweat generation and diagnostic system-status of sweat card	The product is a sweet card which is part of a system intended to measure chloride ions in sweat for the diagnosis of cystic fibrosis. The system runs the sweat test in two stages: iontophoresis sweat stimulation done by delivering pilocarpine through the skin to stimulate sweating (by exciting the sweat glands) and measurement of the chloride concentration in the sweat sample. It includes three components namely: • sweat cards: contains pilocarpine for delivery via iontophoresis and a gel component that collects the resultant sweat; • remote module: sweat card is inserted into a unit, which is placed in direct contact with skin and is held in place via a strap; this battery-powered module stimulates delivery of pilocarpine and resultant production of sweat is collected, analyzed, and sent to a receiver base; • Terminal or receiver base: this contains a unit for calculation and display of results. The manufacturer considered that the sweat card element of the system would be either a Class IIa medical device or an IVD medical device.	The primary purpose of the sweat card is sweat collection; the pilocarpine is acting in an ancillary manner by stimulating glands to ensure collection of sweat is more effective. The sweat card is directly applied to the human body for the purposes of obtaining a sample and therefore is not considered to be an IVD Medical Device as per MEDDEV Guideline 2.14/1 but rather a medical device according to Directive 93/42/EEC. This device incorporates, as an integral part, pilocarpine which, if used separately, can be considered to be a medicinal product, as defined in Article 1 of Directive 2001/83/EC, and that is liable to act on the human body with action ancillary to that of the device. Therefore, the sweat cards should be qualified as medical devices and regarded as Class III medical devices in accordance to Rule 13 in Annex IX of the Medical Devices Directive 93/42/EEC.

	He did not consider the sweat card to be a Class III Rule 13 medical device on the basis that the main purpose of the sweat sensor card is not to elicit a response via a pharmacological action but to diagnose cystic fibrosis by measuring sweat chloride. They considered that the pilocarpine iontophoresis is used for stimulation of sweat as a first step and hence the pilocarpine iontophoresis is an ancillary process to collect sweat.	
Products intended to reduce the effect of alcohol	The product is formulated as a suspension to ingest before drinking alcohol. This kind of suspension is intended to help reduce the absorption of ethylic alcohol in the intestinal lumen after ingestion. The main component of the product is a mineral salt (zeolite). The product is used before the consumption of alcohol with the intention to reduce the blood content of alcohol. Consequently, the product is claimed to reduce the caloric intake, behavioral effects of alcohol, and also the production of substances liable to be toxic linked to the metabolism of alcohol. According to the manufacturer, the product is a medical device because it prevents and reduces the risk of developing some diseases as hepatic cirrhosis, arterial hypertension, ischemic/hemorrhagic ictus.	The reduction of the absorption of ethylic alcohol, without any medical purpose, in order to reduce the blood content in alcohol cannot be considered as a medical purpose. In the case of alcoholism (addictive illness) this product would not have a medical purpose, because it would not help the control of the behavior (compulsive inability to stop drinking), nor of the dependence. The product does not fall within the scope of the definition of medical device set out in Article 1.2 of Directive 93/42/EEC, concerning medical devices, and should not be qualified as such.
Radiation shields	Radiation shields are used to protect against unintended exposure to radiation, for example, when using X-ray machines, during radiotherapy and long-term surgery such as cardiology interventional procedures (occupational exposure). Some radiation shields may be worn by an individual healthcare worker, however other radiation shields may be wall-mounted, mounted on X-ray machinery, ceiling mounted or may be free-standing. Products that are worn by an individual healthcare worker are regulated under the regulations covering personal protective equipment; however, these regulations do not cover products which are not worn by an individual.	The shields are intended to protect persons in the vicinity of the equipment when it is in use (e.g., healthcare professionals) against exposure to radiation-they are not intended to protect the patient. Although the shields will protect against over-exposure to radiation, radiation is not a disease, although it may cause disease. The purpose of these shields is similar to that of filters used for the general protection of the persons in the building. Therefore, these types of radiation shields should not be qualified as medical devices.

	Products intended to protect the patient would be regarded as medical devices. The question is whether radiation shields that are not worn by an individual may be qualified as medical devices.	
Rugby helmet	A helmet that is intended to be worn by any person participating in rugby can protect the skull from non-penetrating direct impacts that could lead to brain injury and other clinical complications.	A helmet worn during rugby can prevent brain tissue damage or other trauma by providing mechanical protection against a hazard, e.g., impact to the head. The helmet does not prevent, treat, or alleviate a disease, nor does it treat or alleviate an injury. Therefore, the helmet does not meet the definition of a medical device and should not be qualified as such.
Autopsy saw	The product under discussion is a saw with extractor unit intended for use only in the mortuary for autopsy. The question has arisen as to whether such a product can be considered to be a medical device, since it is claimed by the manufacturer that it is intended for diagnosis of disease, diagnosis of an injury or handicap or investigation of the anatomy or of a physiological process.	This product should not be qualified as a medical device because it is not intended for use on living human beings.
UV flow germicidal lamp	UV flow germicidal lamps are intended to decrease the level of microbiological load in hospitals (operating rooms, treatment rooms, patient rooms, emergency room, isolation room, and the refuse disposal site). These lamps are able to destroy bacteria, viruses, fungi, and other microorganisms. Contaminated air is drawn by a fan, through a filter that retains dust and other particles, into the disinfection chamber.	Although maintaining clean air may contribute to keeping a patient in an appropriate environment, this is not considered to be a 'medical purpose.' Also, these devices are not intended to be used for disinfecting medical devices but only for disinfecting air. Since these products do not fulfill the definition of a medical device or definition of accessory of medical device, they should not be qualified as such.
Water filter	The product is a water filter with a fiber membrane for removing microorganisms from the water without changing its chemical composition. The manufacturer claims that filtered water is specifically intended for washing wounds and rinsing invasive medical equipment (e.g., endoscopic material) and that the filter should be classified as a medical device of Class llb according to Rule 4 or 15 (both apply according to the manufacturer). The question has arisen as to whether such a product meets the definition of medical device or accessory of medical device.	The water filter does not come into contact with injured skin, and it does not disinfect medical devices, so neither rule is applicable. Water filtration systems that use mechanical barriers such as a fiber membrane for removing microorganisms should be regarded as general hospital equipment and should not be qualified as a medical device or accessory of medical device.

Mattress covers against mites	The product is a type of mattress cover made of cotton fiber fabric which acts as a barrier against mites. Fibers are previously compressed, which, according to the manufacturers, enables the filtration of the finest particles of dust mite, cat, and pollen allergens. The product is placed on the mattress in order to cover it. The user can sleep on the protected mattress, as on any mattress, with the additional use of bed linen. The manufacturers claim that the mattress cover protects the user from allergens in the mattress, and that consequently, the user will be able to use any bed with this protection without onset or progression of the symptoms of the allergy.	A mattress cover does not alleviate the allergic disease. It constitutes only a barrier between the user and a part of its environment, protecting him from allergens. Such a mattress cover could be a part of the treatment plan of patients suffering from allergy by this barrier against dust mite allergens. This fact alone, however, would not qualify the product as a medical device as the mattress cover under bed linen does not act directly on an individual patient. Consequently, a mattress cover against mites should not be qualified as a medical device.
Lubricants intended for body massages and/or sexual intercourse	Lubricants and moisturizers can be effective in relieving pain during intercourse for women with mild to moderate vaginal dryness, or in easing penetration with a condom. According to MEDDEV 2.1/1, the definition of "accessory" requires that an accessory is specifically intended by its manufacturer to be used together with a particular device. The intended use of the accessory must be such as to enable a device to be used in accordance with its intended use. Therefore, a product can only become an accessory to a medical device if the manufacturer of such a product establishes an intended use in conjunction with one or several medical devices.	Water- or silicone-based lubricants that are only intended to be used for body massage or to stimulate sexual intercourse (sexual arousal, pleasure) should not be qualified as medical devices, as such uses do not have a medical purpose. However, water- or silicone-based lubricants which are intended by the manufacturer for the alleviation of vaginal dryness, or for other medical purposes, should be qualified as medical devices. Water- or silicone-based lubricants should be qualified as an accessory to a medical device (condom) if they are necessary for the condom to function as intended, which should be stated by the manufacturer. The fact, or mere indication, that lubricants can be safely used in combination with (latex) condoms does not mean that lubricants enable the condom to be used in accordance with its intended purpose.
Solution for adhesive bandage removal	The product in question is a solution intended by the manufacturer for the removal of adhesive tapes and elastic adhesive bandages. According to the manufacturer, it also removes any adhesive residue which may be left on the skin. It is intended to be applied to the bandage to soak it and acts by dissolving the adhesive, allowing easy removal from the skin.	A solution for removal of adhesive tapes and elastic adhesive bandages, in general, does not fulfill the definition of a medical device, since there is no medical purpose for the solution itself intended by the manufacturer. In general, it also cannot be considered as an accessory to adhesive tapes and elastic adhesive bandages, if those are medical devices, because it is neither intended to enable them to be used in accordance with their intended use (the bandages can be used without the adhesive remover), nor does it directly assist the medical functionality of those bandages in terms of their intended purpose. Nevertheless, some adhesive remover solutions could be considered as medical devices or accessories to medical devices if specifically intended for a medical

		purpose or to enable the correct use of a specific medical device, respectively. An example could be a solution intended to be used with ostomy bags in order to prevent injuries to the stoma if the bags are not removed correctly. In conclusion, a solution for the removal of adhesive tapes and adhesive bandages with no specific medical intended purpose and which cannot be considered as an accessory to a medical device should not be qualified as such.
Pill organizer boxes	Pill organizer boxes are intended for storing the pills in dedicated compartments for a given period for administration at the right time. There are different kinds of boxes: boxes for one day, boxes for one week, 7-day boxes consisting of seven daily compartments and others. The purpose of these boxes is to facilitate and control medication.	Although taking the correct dose of medication at the right time is an important part of medical treatment, tools facilitating this do not fulfill the definition of a medical device, because they are not intended to diagnose, prevent, monitor, treat or alleviate disease themselves. The box is only a convenient tool to help the patient remember whether or not they have taken their medication. Since these pill organizer boxes do not fulfill the definition of a medical device, they should not be qualified as such.

10.10. BORDERLINE *IN VITRO* DIAGNOSTIC (IVD) MEDICAL DEVICES

Let us recap some definitions according to the Regulations:

> The definition of *in vitro* diagnostic medical devices reads as follows:
> Article 1(2) b of the IVD Directive 98/79/EC *"'in vitro diagnostic medical device' means any medical device which is a reagent, reagent product, calibrator, control material, kit, instrument, apparatus, equipment, or system, whether used alone or in combination, intended by the manufacturer to be used in vitro for the examination of specimens, including blood and tissue donations, derived from the human body, solely or principally for the purpose of providing information:*
>
> - *concerning a physiological or pathological state, or*
> - *concerning a congenital abnormality, or*
> - *o determine the safety and compatibility with potential recipients, or*
> - *to monitor therapeutic measures."*

From this definition it follows that in order to fall within the definition of an IVD medical device, the product must also meet the definition of a medical device.

Article 1(2)a of the MD Directive 93/42/EEC *"'medical device' means any instrument, apparatus, appliance, material or other article, whether used alone or in combination, including the software necessary for its proper application intended by the manufacturer to be used for human beings for the purpose of:*

- *diagnosis, prevention, monitoring, treatment, or alleviation of disease;*
- *diagnosis, monitoring, treatment, alleviation of or compensation for an injury or handicap;*
- *investigation, replacement, or modification of the anatomy or of a physiological process;*
- *control of conception, and which does not achieve its principal intended action in or on the human body by pharmacological, immunological or metabolic means, but which may be assisted in its function by such means."*

Article 1(2) b IVDD "*Specimen receptacles are considered to be in vitro diagnostic medical devices. 'Specimen receptacles' are those devices, whether vacuum-type or not, specifically intended by their manufacturers for the primary containment and preservation of specimens derived from the human body for the purpose of in vitro diagnostic examination.*"
Article 1(2)b IVDD *"Products for general laboratory use are not in vitro diagnostic medical devices unless such products, in view of their characteristics, are specifically intended by their manufacturer to be used for in vitro diagnostic examination."*

Examples as given by the Borderline Manual are discussed further below:

Borderline Device	Background	Outcome
Sample receptacles and sampling devices which are intended to be used for the collection by the lay user of samples, which are subsequently examined by third persons.	The products in question are IVD kits which are being supplied to the public for a variety of medical conditions, including tests for food allergies and infections such as Chlamydia. These tests are being supplied by post to members of the public in the following manner: The patient orders the kit from the company concerned and the kit is dispatched to him. The kit contains the required equipment to take a sample. The patient is instructed to take the sample (for example, usually a urine sample or blood sample-either via a lancet, or they are advised to take their kit to their doctor for a blood sample to be taken). The sample is then placed in some type of storage container. Once the sample is obtained, the patient is instructed to send it back to the company supplying the kit. The patient is then supplied with the result of the test, indicating whether or not they have a positive result. None of these activities involve healthcare professionals.	These kits are *in vitro* diagnostic medical devices by means of applying Article 1(2) b IVDD, which states that: "Specimen receptacles are considered to be *in vitro* diagnostic medical devices. 'Specimen receptacles' are those devices, whether vacuum-type or not, specifically intended by their manufacturers for the primary containment and preservation of specimens derived from the human body for the purpose of *in vitro* diagnostic examination." The question arises whether these specimen receptacles could be considered as a 'device for self-testing' in accordance with Article 1(2) d IVDD according to which 'device for self-testing' means any device intended by the manufacturer to be able to be used by laypersons in a home environment; In this determination, the notion 'used' is essential. Firstly, it is necessary to examine the instructions for use. Where the instructions for use require an action to be taken by the end-user of the device in question, the notion 'used' is fulfilled. In addition, the definition of 'self-testing' provides guidance on the action to be taken, i.e., "testing." Thus, where a specimen receptacle is simply used by the patient to contain a specimen it remains a specimen receptacle. To become a 'device for self-testing,' either the filling of the receptacle with a specimen should result directly in a result being given or the patient should need to do something directly to the specimen prior to its dispatch in order to fulfill the concept of 'used.'

<table>
<tr>
<td>CE labeled microscope slides</td>
<td colspan="2">This product is a microscope slide which is made of a thin sheet of glass used to hold objects for examination under a microscope. Unless the manufacturer's intended purpose falls within the definition of an in vitro diagnostic medical device, it must be regarded as a general laboratory product. The latter are excluded from the IVDD by Article 1(2) b IVDD.
In addition, MEDDEV 2.14/1 rev.1 states in this context:
"If, however, the product does not in fact possess specific characteristics that make it suitable for one or more identified in vitro diagnostic examination procedures, then the manufacturer is not free to bring it within the scope of the IVDD merely by affixing the CE marking to it. In other words, a manufacturer is not able to bring within the scope of the IVDD a product that, in reality, is a piece of general laboratory equipment simply by affixing the CE mark to it."</td>
</tr>
<tr>
<td>Single or multiple channel pipettes</td>
<td colspan="2">The single or multiple channel pipettes are used for aspirating and dispensing specific volumes in the microliter scale. The volume is set by rotating the thumbwheel or the push-button. These pipettes have various laboratory purposes.
Unless the manufacturer's intended purpose falls within the definition of an in vitro diagnostic medical device, these pipettes must be regarded as a general laboratory product. The latter is excluded from the IVDD by Article 1(2) b IVDD.
In addition, MEDDEV 2.14/1 rev.1 states in this context:
"If, however, the product does not in fact possess specific characteristics that make it suitable for one or more identified in vitro diagnostic examination procedures, then the manufacturer is not free to bring it within the scope of the IVDD merely by affixing the CE marking to it. In other words, a manufacturer is not able to bring within the scope of the IVDD a product that, in reality, is a piece of general laboratory equipment simply by affixing the CE mark to it."
Moreover, the MEDDEV 2.14/1 rev. 1 specifically refers to pipette. Under point 4 "Products for general laboratory use," it is mentioned that pipettes are laboratory products that are not usually considered to fall within the scope of the IVD directive.</td>
</tr>
<tr>
<td>Qualification of fluid collection bowl</td>
<td>The product is a single-use plastic device consisting of a round container, a cover, a drainer, and a liquid-absorbent inner layer. The absorbent material would receive the waste fluid from syringes, therefore avoiding spillage.
This product is intended to facilitate the safe disposal of fluids during a medical procedure (e.g., in operating theaters), avoiding any spills or splashes.</td>
<td>The product's intended purpose is the safe disposal of body fluids collected during a medical procedure, and therefore this product is not intended to be used so as to achieve a medical purpose.
This bowl is similar to containers for used syringes, needles, and other bio-hazardous materials or waste, which are not qualified as medical devices. They could also be compared to gallipots, which, as stated in the entry 1.10. of the Manual on Borderline and Classification, should not be qualified as medical devices.
Therefore, this product should not be qualified as a medical device.</td>
</tr>
</table>

10.11. BORDERLINE ACTIVE IMPLANTABLE MEDICAL DEVICE

Borderline Device	Background	Outcome
Bone anchored hearing aids	Bone anchored hearing aids consist of a titanium implant (that is screwed into the patient's cranium), an abutment to transfer vibrations and a sound processor (hearing aid). The sound processor collects sound, amplified the vibrations, and passes them on to the implant (via the abutment). The vibrations find their way to the inner ear without the help from any other devices and make it possible for the patients to hear. The sound processor relies on an electrical power source, while both the abutment and the implantable titanium piece are passive. It must be noted that bone-anchored hearing aids are different from cochlear implants, which are intended to stimulate the hearing nerve directly via an electrode array inserted into the cochlea (inner ear). According to MEDDEV 2. 1/2 rev 2, cochlear implants activated by an external power transmitter are regarded as active implantable medical devices as the implanted component clearly depends on a power source for its function and its purpose is to convert the power it receives into electri	Guidance MEDDEV 2. 1/2 rev 2 states that a product falls under the definition of an active implantable medical device if it is at the same time both active and implantable. Under its point 2.1.2 the MEDDEV 2.1/2 rev.2 excludes that a device is active if the function of the device is only a mere transmission of vibration. With the bone-anchored hearing aid, the implantable component is not an active medical device-it is a titanium 'rod.' The sound processor, which is the active component, is not implanted. On the basis of the above, bone-anchored hearing aids fall under Directive 93/42/EEC on medical devices. The sound processor shall be classified as Class IIa medical device according to classification Rule 9, while the implant is regarded as Class IIb medical device according to classification Rule 8 of Annex IX of Directive 93/42/EEC. Where the two elements are supplied as one system, the higher classification applies to the overall product, and the system is therefore classified as Class IIb medical device.

	cal signals which trigger appropriate sensory channels in the brain. The question is whether bone-anchored hearing aids fall under the Directive 93/42/EEC on medical devices or under the Directive 90/385/EEC on active implantable medical devices.	

10.12. BORDERLINE MEDICAL DEVICE-MEDICINAL PRODUCT

Let us recap the definitions according to the regulations:

Medical device definition (Article 1(2)a of Directive 93/42/EEC, as amended):

(a) 'medical device' means any instrument, apparatus, appliance, material, or other article, whether used alone or in combination, including the software necessary for its proper application intended by the manufacturer to be used for human beings for the purpose of:

- *diagnosis, prevention, monitoring, treatment, or alleviation of disease;*
- *diagnosis, monitoring, treatment, alleviation of or compensation for an injury or handicap;*
- *investigation, replacement, or modification of the anatomy or of a physiological process;*
- *control of conception; and*
- *which does not achieve its principal intended action in or on the human body by pharmacological, immunological, or metabolic means, but which may be assisted in its function by such means.*

Medicinal product definition (Article 1(2) of Directive 2001/83/EC, as amended):

a. *Any substance or combination of substances presented as having properties for treating or preventing disease in human beings; or*
b. *Any substance or combination of substances which may be used in or administered to human beings either with a view to restoring, correcting, or modifying physiological functions by exerting a pharmacological, immunological, or metabolic action, or to making a medical diagnosis.*

Examples as given by the manual are discussed below:

Borderline Device	Background	Outcome
Product for testing patient reflex cough	The product may be used, following initial diagnosis, by health care professionals as a part of a neurological evaluation to assess the patient's laryngeal cough reflex (LCR), neurological airway protection and the vagal (cranial nerve X) component of reflex cough. The test uses 1–3 inhalations of a nebulized 20% solution of L-(+)-tartaric acid in a medicinal non-ventilatory nebulizer. Tartaric acid acts as a mild irritant to the mucosa of the larynx. The resultant involuntary reflex response, i.e., cough, is indicative of whether the upper airway is neurologically protected. The expected result of a normal test is an immediate series of a forceful coughs, which are primarily expiratory "airway cleaning" in character. The expected result of an abnormal test is represented by an absence of coughing, or a diminished (weak) coughing, or coughing not immediately after administration of the test stimulus. The testing is terminated when the subject either elicited a cough or failed to cough after three valid inhalations. According to the manufacturer's statements the test functions like many other devices which assess neurological pathways, including the reflex hammer, nerve conduction stimulators, and the fiberoptic endoscopic evaluation of the	The tartaric acid is intended to be used for neurologically impaired patients to test the functioning of their laryngeal cough reflex in order to determine appropriate treatment. A pharmacological action on the patient cannot be excluded. The tartaric acid is used as an irritant to produce a cough reflex and is a substance administered for *in-vivo* diagnostic purposes. The administration of this substance could be considered as a component of the general medical diagnosis since it consents to improve definition of the extent of neurological damage. On the basis of the above, the product does not meet the definition of a medical device.
	swallowing with sensory testing, where a physical stimulus (either electrical or physical) challenges a neurological pathway to elicit a response. The manufacturer regards 20% tartaric acid (TA) solution used to induce cough as a physical stimulus, although it cannot be denied that it induces chemical changes to the sensory nerves and that there may be "pharmacological" agents involved. Another consideration given by the manufacturer is that TA in the dose used will have no measurable effect on general body metabolism, and it should therefore not be considered to have a "metabolic" action. Finally, the manufacturer stresses that the purpose of the TA test is not diagnostic. As with the use of other physical stimuli, the patient's diagnosis will already have been established (e.g., stroke) and the test will be used to determine appropriate treatment for the individual patient. solution.	

	The item will be used with a commercially available disposable jet nebulizer for single-patient use (the nebulizer is not pre-filled with the solution). Continuous flow will deliver the agent through the inspiratory cycle, and will allow re-charge of the agent during the exhalation phase. According to the manufacturer's statements, the use of the solution is deemed ancillary to the nebulizer, as the solution cannot be administered without nebulization.	
Elastoviscous fluids	This product is a sterile, non-pyrogenic, elastoviscous fluid containing Hyland (derivatives of hyaluronan-sodium salt of hyaluronic acid that consists of repeating disaccharide units of N-acetylglucosamine and sodium gluconate). It is biologically similar to hyaluronan, which is a component of synovial fluid which is responsible for its viscoelasticity. The product achieves its therapeutic effect through viscosupplementation, a process whereby the physiological and rheological states of the arthritic joint tissues are restored. Viscosupplementation is a treatment to decrease pain and discomfort, allowing more extensive movement of the joint. The product has to be injected into the affected joint to achieve its effect. Hylands are degraded in the body by the same pathway as hyaluronan.	The data presented does not allow issuing a general statement for the qualification of the elastoviscous fluids. Viscoelastic materials with intended use for mechanical/physical purposes such as protection of tissues during and after surgery and separation of tissues are considered to be medical devices. Such materials are also used as synovial fluid replacements where viscosupplementation provides support and lubrication. Additional pharmacological benefits claimed which are ancillary to the mechanical action do not alter the medical device status. However, certain of these materials, such as some hyaluranon based products, where the predominant claims are of a pharmacological nature and not primarily related to any viscoelastic characteristics, are classed as medicinal products. Therefore, it is appropriate to follow a case-by-case approach, taking into account all product characteristics and in particular:
		• The intended purpose of the product and the claims of the manufacturer, taking into account the way the product is presented; • The nature of the principal intended action. For example, is there any pharmacological or metabolic action (e.g., anti-inflammatory effect, stimulation of *in vivo* hyaluronic acid synthesis, etc.), not ancillary to the mechanical/physical action of the product?

In-vitro fertilization (IVF) and assisted reproductive technologies (ART) products	The *in-vitro* fertilization (IVF) and Assisted Reproductive Technologies (ART) products cover a large spectrum of products. For some of these products the principal intended action is clearly a pure physical or mechanical action. On the other hand, a number of products that fall within this category contain substances that act by pharmacological, immunological, or metabolic action. In the latter cases, it is of utmost importance to assess whether such a pharmacological, immunological, or metabolic action represents an ancillary or primary action. It is concluded that this analysis should be done on a case-by-case basis.	IVF/ART products may be qualified and regulated as medical devices provided that they meet the definition of a medical device as laid out in Directive 93/42/EEC, taking into consideration the principal intended action and intended purpose of the product. The concept of 'used for human beings' is interpreted in the broadest sense. The whole IVF/ART procedure and related products would be seen as (indirectly) "(…) used for human beings for the purpose of (…) replacement or modification of (…) a physiological process" by promulgating pregnancy. Therefore, the definition of medical devices can include IVF/ART products. Examples of products which could be qualified as medical devices (classification is only indicative and must be assessed on a case-by-case basis taking into account all product characteristics): • Devices that act in a physical or mechanical way intended to be used for IVF/ART (such as pipettes or syringes) should be classified according to the rules set out in Annex IX of Directive 93/42/EEC, depending mainly on their intended use; Devices, such as washing, separating, sperm immobilizing, cryoprotecting solutions, which are liable to act with close contact on the inner or outer cells during the IVF/ART are likely to be considered as Class IIb medical devices, in particular by analogy of Rule 3. • Devices manufactured utilizing animal tissues or derivatives rendered non-viable are considered as Class III medical devices according to Rule 17; o Devices incorporating, as an integral part: o a human blood derivative or o a substance which, if used separately, can

		be considered to be a medicinal product, as defined in Article 1 of Directive 2001/83/EC, and which is liable to act on the human body with action ancillary to that of the devices, are considered as Class III medical devices according to Rule 13. The assessment of the ancillary nature of the pharmacological, immunological, or metabolic action of any medicinal product contained in IVF/ART products should be done on a case-by-case basis, taking also into account the purpose of the inclusion of this substance into the product.
Peritoneal dialysis solutions	Solutions for peritoneal dialysis are preparations for intraperitoneal use which contain electrolytes in a similar concentration to that in plasma, and also contain glucose or another suitable osmotic agent. Peritoneal dialysis solutions always contain sodium, chloride, and hydrogen carbonate or a precursor. They may also contain calcium, magnesium, and potassium. In peritoneal dialysis, the solution is infused into the peritoneal cavity, where exchange of electrolytes takes place by diffusion and convection, and excess fluid is removed by osmosis, using the peritoneal membrane as an osmotic membrane. Such exchange of electrolytes induces a metabolic effect.	Peritoneal dialysis solutions are used for specific and restricted medical conditions to be administered parenterally to patients with an identified medically diagnosed condition and have a metabolic mode of action. Therefore, such solutions cannot be qualified as medical devices. -Devices manufactured utilizing animal tissues or derivatives rendered non-viable are considered as Class III medical devices according to Rule 17; • Devices incorporating, as an integral part: o a human blood derivative or o a substance which, if used separately, can be considered to be a medicinal product, as defined in Article 1 of Directive 2001/83/EC, and which is liable to act on the human body with action ancillary to that of the devices, are considered as Class III medical devices according to Rule 13. The assessment of the ancillary nature of the pharmacological, immunological, or metabolic action of any medicinal product contained in agents for transport, nutrition, and storage of organs intended for transplantation should be done on a case-by-case basis, taking also into account the purpose of the inclusion of this substance into the product.

		In accordance with Article 2(2) of Directive 2001/83/EC, in case of doubt we're taking into account all product characteristics, and provided that the concerned product meets both definitions of a medicinal product and of a medical device, the provisions of Directive 2001/83/EC shall apply.
Zinc oxide containing creams	Products containing zinc oxide are available as creams for local administration. Some zinc oxide containing products, depending on their claims, and intended use, might be covered by Directive 76/768/EEC on cosmetic products. The discussion below only concerns zinc oxide containing products used to treat or prevent minor skin irritations (e.g., burns, cuts, nappy rash, eczema, etc.).	A medical device should not achieve its principal intended action in or on the human body by pharmacological, immunological, or metabolic means, but maybe assisted in its function by such means. For zinc oxide containing products, according to the literature, a pharmacological and metabolic action is demonstrated, e.g., may play a role in enzymatic processes, support of wound granulation. The pharmacological action may, however, be ancillary when the product concerned is primarily a barrier cream. In such cases, the qualification of zinc-oxide containing products is defined taking into account the claims, the intended purpose, and the relevant primary mode of action. Some products that act primarily as a barrier may therefore be acceptable as Class III medical devices in accordance with Rule 13 of Annex IX of Directive 93/42/EEC. In accordance with Article 2(2) of Directive 2001/83/EC, in case of doubt we're taking into account all product characteristics, and provided that the concerned product meets both definitions of a medicinal product and of a medical device, the provisions of Directive 2001/83/EC shall apply.

Eye drops intended for related to the alleviation of 'soreness,' 'discomfort' or 'irritation' caused by environmental factors (such as dust, heat, smoke, etc.)	On the EU market, there are many different types of eye drops used for different purposes. • Eye drops regulated as medicinal products: Eye drops used in or administered to human beings with a view to making a medical diagnosis are regulated as medicinal products in accordance with Article 1(2) of Directive 2001/83/EC. Eye drops with pharmacological, immunological, or metabolic principal mode of action will fall under the definition of a medicinal product if they may be used in or administered to human beings either with a view to restoring, correcting, or modifying physiological functions. • Eye drops regulated as medical devices: Products specifically intended to be used for disinfecting, cleaning, rinsing or, when appropriate, hydrating contact lenses are medical devices. Eye drops with a physical mode of action that are clearly indicated for a medical purpose are acceptable as medical devices (e.g., for the treatment of hay fever). Therefore, the discussion below only concerns products for which no 'medical claims are made, nor any claims associated with the use of contact lenses. Commonly the only claims made relate to the alleviation of 'soreness,' 'discomfort' or 'irritation' caused by environmental factors (such as dust, heat, smoke, etc.), or simply 'for tired eyes.'	The qualification of a particular eye drop products shall depend upon the intended purpose and mode of action of the product and must be assessed on a case-by-case basis. Some products may claim indications such as 'treatment' of a physiological response to external factors (i.e., dust, smoke, dry heat, air conditioning) and are used to create a humidification of the eyes or to treat minor irritations of the eyes. As such, they are designed to complement physiological functions provided by normal tears to help meet an increased demand for humidification/lubrication of the eyes. These products do not need to stimulate the physiological functions of the natural tears. The intended effect may be achieved by physical means only, i.e., by washing the eye surface (effect of dust, smoke), or supplement the aqueous layer of natural tear film with additional water under conditions of significantly increased evaporation (dry heat, air conditioning). By decreasing the exposure to irritants, they may be helpful in the treatment of 'minor irritations' of the eye. Therefore, products which claim to treat damage caused by environmental factors may be acceptable as medical devices (or medicinal products, dependent on their mode of action) provided that they clearly claim to treat or alleviate damage to or irritation of the eye (e.g., via repairing damage to the tear film caused by environmental conditions). These products must not claim to be artificial tears or any kind of replacement of natural tears. Products called 'artificial tears' are usually those that replace or complement natural tear functions when natural tear function has been compromised. The term 'artificial tears' should therefore only be used for products that replace the function of natural tears. Such products may be regulated as either medical devices or medicinal products, depending upon their mode of action.

Product for use in acute sore throat	The product is intended to be used, according to the instruction for use, in acute sore throat with irritated mucosa. This product contains Icelandic Moss and benzocaine. The mechanism of action of Icelandic Moss is a mucilaginous drug which spreads on the oral and pharyngeal mucosa, forming an internal coating on the mucosa, acts by covering the mucosa with a protective layer. The benzocaine helps reducing further sensitivity. The European Pharmacopoeia contains monographs for "Iceland Moss" and "Benzocaine (benzocaine)." Icelandic moss may be considered as a medicinal substance and is described as having pharmacological properties including anti-inflammatory, antibacterial, and antiviral activities. Benzocaine is a known medicinal substance with pharmacological action as local anesthetic and is found in several medicinal products intended for temporary local relief of pain associated with sore throat.	A pharmacological action of Icelandic moss cannot be excluded, as it is described to have pharmacological properties including anti-inflammatory, antibacterial, and antiviral activity. Furthermore, even if the Icelandic moss may be considered as providing a physical barrier, the benzocaine contained in this product could not be considered having an ancillary action. On the basis of the above, it is concluded that this product achieves its principal intended action by pharmacological means and therefore do not meet the definition of a medical device.
Plaster with capsaicin	The products in question are warming plasters (adhesive) containing capsaicin (capsicum oleoresin or capsicum extract). These plasters are intended for heat-treatment, local analgesia and are indicated to treat muscular, rheumatic, and neuralgic pains. The European Pharmacopoeia contains monographs for "Capsici fructus" and "Capsicum Oleoresin." The European Scientific Cooperation on Phytotherapy (ESCOP) has classified "Capsici fructus" as an herbal medicinal product. According to classification Rule 13, devices incorporating, as an integral part, a substance which, if used separately, can be considered to be a medicinal product, and which is liable to act on the human body with action ancillary to that of the device, are in Class III.	The adhesive plaster described appears to be acting as a carrier for the substance (Capsaicin) in order for the substance to be delivered to the body to produce the analgesic effect. It therefore cannot be excluded that this product achieves its principal intended action by pharmacological means. Therefore, warming plasters with capsaicin may not be qualified as a medical device.

Gold implants for treatment of osteoarthrosis	Various gold salts are available for systemic treatment of painful joint diseases, mainly arthritis and arthrosis. There is increasing evidence that the gold salts exert their effect by a direct intervention with the cellular immune response. It is generally assumed that the active part of the gold salts is the gold ions. Solid gold implants for local treatment of painful joint diseases in animals have been used by veterinarians for a number of years. Several studies have documented a therapeutic effect of solid gold implants in animals. Research has provided evidence that gold ions are released from the implanted gold and diffuse into the surrounding tissue, where they can be detected in connection with inflammatory cells from the immune system. The researchers concluded that solid gold implants as a local treatment, mimics systemic treatment with a gold-containing drug. The question is whether gold implants should be qualified as medical devices or not.	In the view of the above-mentioned mode of action which mimics systemic treatment with a gold-containing drug, gold implants for treatment of osteoarthrosis exert their effect by a direct intervention with the cellular immune response and therefore cannot be qualified as medical devices.
Substances for chemical peeling	Skin exfoliation is widely used to reduce the signs of aging in the skin since it diminishes imperfections by peeling away the skin's top layers (solution to sun-damaged, unevenly pigmented, and finely wrinkled facial areas). It is also used on acneic skin and on acne scars. Skin exfoliation could be obtained by different means such as mechanical peeling (also called dermabrasion which consists of a sandblasting technique with particles), physical peeling (carbon dioxide ultra-pulsed laser, erbium: YAG laser) and chemical substances. In this case, the depth of peeling depends of many factors such as the substance used (e.g., alpha hydroxy acids (AHA), beta hydroxyl acids as salicylic acid, trichloroacetic acid (TCA) and phenol), its concentration, the pH of the solution, and the length of the application. Considering their intended use on acneic skin, some manufacturers requested an opinion about placing chemical peeling containing salicylic acid (and glycolic acid) as Class I medical device. The question is whether these chemical peeling must have a specific action on treatment of acne to be considered as a medical device or whether the claim that they are indicated in case of acneic skin could be sufficient to qualified them as medical device.	It is concluded that, depending on their principal mode of action and in the case, they have a specific medical purpose (for example, treatment of acne), these products might be qualified as medical devices. However, the claim that they may be used on acneic skin is not considered sufficient to indicate a medical purpose. When these products could be qualified as medical devices, they shall be classified as Class III medical devices according to Rule 13 if they contain a medicinal substance with ancillary action.

Mustard packs	Mustard packs, consisting of mustard seed powder, are usually applied on the chest, abdomen, or articulations and the intended use is to alleviate the symptoms of common colds and other respiratory system ailments, and rheumatism. The mechanism of action of the mustard pack remains unclear, and the properties which could prove their suitability for the treatment of disease and prophylaxis are still unknown. The mustard packs work in a manner analogous with Capsaicin plasters (see Section 4.9.).	It is considered that the mustard seed powder produces a heating effect by causing capillaries to expand and an analgesic effect by interacting with skin receptors to reduce sensitivity due to the release of allyl isothiocyanate compounds that follows the hydrolyzation of the glucosinolates contained in the seeds. Therefore, mustards packs principal intended action is achieved by pharmacological means and cannot be qualified as medical devices.
Washing solutions used for pathogenic microorganisms	There are washing solutions with antimicrobial properties, destined for the disinfection of healthy skin, being placed on the market as Class III medical devices, even though, according to European guidance, topical disinfectants (antiseptics) for use on patients are given as examples of medicinal products. According to manufacturer information, these washing solutions (which are presented as concentrate solutions, gel, foam, and impregnated wipes) are intended for multi-drug resistant organisms' decolonization by physical cleansing (inhibiting their spreading and transmission with sustained anti-microbial barrier effect), and may also be used for pre-operative cleansing. General treatment indications include washing of hair, face, ears, nose, upper body, lower body, perineum, urethra, and catheter entry sites, legs, and others such as spectacles, jewelry, and prosthesis. These washing solutions are also presented as oral solutions/mouthwashes for: the disinfection of the oral cavity and pharynx; prevention of plaque formation, caries, periodontitis, gingivitis, and are applicable on impaired mucous membranes, according to manufacturer claims.	Considering that: • These washing solutions for inhibition of pathogen microorganisms present medical claims such as: multi-drug resistant organisms' decolonization, prevention of plaque formation, caries, periodontitis, and gingivitis; • Only products intended for disinfecting medical devices or cleaning contact lenses (Rule 15 of Annex IX of the MMD) are medical devices; • As indicated on point 5.1 of the Manual on Borderline and Classification: "Hand disinfectants do not appear to be qualified as an accessory to a medical device. These products are for disinfecting the hands and not devices;" • According to MEDDEV 2.1/3 rev 3, Section A.2.2.2.: "Topical disinfectants (antiseptics) for use on patients" are given as examples of medicinal products; -Mouthwashes are medicinal products when the intended purpose is to treat or prevent oropharynx diseases; • The solutions presented have a pharmacological main mode of action as antiseptics/disinfectants. Therefore, washing solutions used for pathogenic microorganisms cannot be considered to be medical devices.

Mousse for rapid relief from irritation, itching, burning, and sensitivity associated with chickenpox	A product, available as a mousse (foam), designed to relieve the symptoms of chickenpox on large surfaces of the skin and used for rapid relief from irritation, itching, burning, and sensitivity associated with chickenpox, was placed on the market as a Class I medical device, according to Rule 4 of Annex IX of the Directive 93/42/EEC. According to information provided by the manufacturer, the product is based on a patented complex (that results from the filtration process of the Aloe barbadensis gel) which helps to protect and soothe the irritated skin, this ingredient supports the skin's own natural immune system. Its polysaccharide chains have the unique characteristic of binding to harmful bacteria. This blocks the bacteria, thus protecting the skin cells from being attacked. This patented complex creates a barrier for pathogenic bacteria avoiding the invasion of the hosts' tissue.	Considering that: • According to expert opinion, this complex acts in a manner similar to that of the body's immunological response; E.g., There are several receptors for the recognition of pathogenic organisms, some of them are present in the bloodstream and in tissue fluids as soluble blood serum proteins, and others can be found in cell membranes of macrophages, neutrophils, and dendritic cells. An example of soluble forms that bind to the microbial surface and promote their opsonization is mannose-binding lectin and C-reactive protein. These receptors, when connected to the microbial surface, also have the ability to trigger the complement system, turning the invader into a sort of target for the complement-mediated lysis. • There is an antimicrobial action through the blockage of pathogenic bacteria's receptors; • There is insufficient data to demonstrate a mere physical action; -In case of doubt we're taking into account all product characteristics, and provided that the concerned product meets both definitions of a medicinal product and of a medical device, Article 2(2) of Directive 2001/83/EC could apply. On the basis of the above, it is concluded that this product does not meet the definition of a medical device.

Injectable substance for treatment of localized adiposity	The product is described by the manufacturer as an injectable aqueous solution with a micro-gelatinous base, which helps to modulate and favor the action of external ultrasound waves (generated by the use of a medical ultrasound generator) for the non-surgical treatment of localized adiposity by means of intralipotherapy. It is placed on the market as a Class III medical device. According to manufacturer information, the product is a totally resorbable solution, composed of an aqueous gelatinous micro-containing polymer. The solution is injected into the adipose tissue, during intralipotherapy technique, and produces a swelling in the tissue, and, when followed by the external application of ultrasound to the skin, leads to the generation of micro-cavitation in the area where the solution was injected, increasing the drainage of the fluid present in the treated area. Also, according to the labeling and instructions for use, this product has as the intended use simply the treatment of localized adiposity, without any relation to a specific pathology, although it bears the instructions that it is intended to be used only by physicians. The IFU mention the use on the "culotte de cheval" as well as in medicine and esthetic surgery/dermatology, as the manufacturer intends the product to be used for the treatment and improvement of body imperfections due to small accumulations of adiposity. The indication is limited only to healthy patients.	Considering that the treatment of localized adiposity on healthy patients is not a medical purpose and that the product does not have any medical claim, it is concluded that the product does not meet the definition of a medical device and cannot be qualified as such.

Riboflavin solution for treatment of keratoconus	The product is a solution containing riboflavin supplied in the form of eye drops and is intended for the treatment of keratoconus (a degenerative disorder of the eye which causes structural changes within the cornea). The riboflavin solution is administered into the eye and is activated via illumination with UV-A light for approximately 30 minutes. The riboflavin causes new bonds to form across adjacent collagen strands in the stromal layer of the cornea, which increases the tensile strength of the cornea. The intended purpose of the product is to increase the collagen cross-linking by using riboflavin in the treatment of keratoconus by causing the collagen fibrils to thicken, stiffen, and cross-link and reattach to each other making the cornea stronger, more stable and in turn halting the disease progression. The question is if the mode of action of the riboflavin in this clinical indication is a chemical reaction or a pharmacological, immunological, or metabolic action.	The available information indicates that the riboflavin has a dual function, firstly on the production of oxygen free radicals, and secondly by absorbing the UV-A radiation and preventing damage to deeper ocular structures, such as corneal endothelium, the lens, and the retina. The application of riboflavin results in an alteration of the normal chemical process of cross-linking of collagen. Considering this, as well as the definition of medical device and metabolic action included in MEDDEV 2. 1/3 rev 36 (an action which involves an alteration, including stopping, starting, or changing the speed of the normal chemical processes participating in, and available for, normal body function), it can be concluded that this product should not be qualified as a medical device.
Dentistry products with aluminum chloride used in hemostasia	Dentistry products with aluminum chloride are used in hemostasia. These products contain aluminum chloride in various concentrations. Liquids and gels contain from 20% to 25% aluminum chloride, while impregnated retraction cords contain from 5% to 10% The products formulated as liquid and gel are intended to staunching perigingival bleeding that results from decay cavities preparation. The aluminum chloride provides a local astringent effect. The action of these products is based on precipitation of albumins which in turn block the vessels (capillaries). These products are used on the mucous membranes or injured skin, creating a protective layer and contracting gums. It is claimed that bleeding stops after several minutes enabling single day treatment without the need for temporary dressing. The impregnated retraction cord is used for retraction of the gingival tissues around the teeth for improving the results of dental impressions and hemostasis of the gingival margin. The aluminum chloride reduces the liquid in the gingival pocket and closes the smaller blood vessels. The effective retraction when the cord is correctly placed would take few minutes. Some product types additionally contain lidocaine.	As that the mode of action of aluminum chloride is other than pharmacological, immunological, or metabolic, these products should be qualified as medical devices.

Qualification and classification of a wound gel containing soluble beta-glucan	The product is a sterile, non-preserved, amorphous, and thixotropic wound filling gel containing water, glycerol, carboxymethylcellulose, and soluble beta-glucan (SBG) 2%. It is to be administered to the wound surface as a primary wound dressing, which then would be covered with a conventional secondary dressing. The question is whether such a product would be qualified as a medical device and classified as a Class III medical device with ancillary medicinal action under classification Rule 13. The soluble beta-glucan utilized is an aqueous soluble 1,3/1,6-glucan produced from baker's yeast; it is a polymer containing glucose as the only monomer. A 2% formulation in water presents as an adhesive gel-like solution. SBG has been found to be an inducer of innate immune responses through its action on phagocytic cells at the beta-glucan receptor Dectin-1, which stimulates macrophage migration into wounds. The gel product also contains carboxymethylcellulose (CMC) and glycerol, which are both established components of wound healing dressings presented as gels. The manufacturer claims that the product has as its primary function, as a hydrogel wound dressing, the moistening, and an assisting debriding effect in the wound bed. This is achieved by the functionality of glycerol and CMC being able to bind excessive amounts of water (i.e., having a high viscosity even at 1 to 2% concentration), but also by the soluble beta-glucan polysaccharide component which forms a weak gel at a 2% concentration in water. The soluble beta-glucan is claimed to have a dual-action. The ancillary, or medicinal substance effect, is claimed to stimulate the wound healing process by modulating phagocytic cells in the wound bed, especially macrophage functions. Soluble beta-glucan may also contribute to reducing the risk for secondary wound infection by modulating the functions of cells belonging to the innate immunological system and by recruiting phagocytic cells into the wound area.	For this specific product, the data provided sufficient evidence to indicate that while the principal action is exerted by the moist wound healing effect, the soluble beta-glucan which, if used separately, can be considered to be a medicinal product, as defined in Article 1 of Directive 2001/83/EC, is liable to act on the human body with action ancillary to that of the wound healing process. Therefore, this product should be qualified as a medical device and regarded as a Class III medical device under Rule 13. This should not be taken to mean, however, that other products containing soluble beta-glucans or other dual-action products should always be qualified as medical devices.

Glycerin suppositories	This product is used to relieve occasional constipation. Glycerin suppositories act as a hyperosmotic laxative by drawing water into the intestine. This effect usually results in a bowel movement. Glycerin suppositories act by increasing the transportation of electrolytes through the bowel membrane (osmotic effect) and by stimulating nerve endings of rectum triggering defecation reflex (stimulant effect). In this case an osmotic process leads to a metabolic effect while the stimulant effect is considered as a pharmacological mode of action.	Glycerin suppositories are a product intended to treat and prevent constipation that act via a metabolic and pharmacological mode of action and should not be qualified as a medical device.
D-mannose for the prevention of urinary tract infections	D-mannose for the prevention of urinary tract infections (UTIs) was placed on the European market in 2015 as a Class IIa medical device. UTIs are predominantly caused by Escherichia coli (E. coli). The first step in infection is the adhesion of E. coli to the urothelium, the epithelial cells of the urinary tract. This step is mediated by the binding of FimH-a mannose-binding protein located at the tip of bacterial type 1 fimbriae to the mannosylated proteins, called uroplakins, coating the urothelium. *In vitro* experiments have shown that D-mannose binds to FimH, thereby preventing the binding of E. coli to the uroplakins. At a sufficient concentration in urine, D-mannose causes saturation of FimH and prevents E. coli from colonizing the urothelium. At the same time, the reduction of bacterial levels in urine by D-mannose has also been confirmed in *in vivo* animal UTI models.	According to the judgment of the Court of 6 September 2012, case C-308/11, Article 1(2)(b) of Directive 2001/83/EC must be interpreted as meaning that, for a substance to be regarded as exerting a "pharmacological action" within the meaning of that provision, it is not necessary for there to be an interaction between the molecules of which it consists and a cellular constituent of the user's body, as an interaction between that substance and any cellular constituent present within the user's body may be sufficient. Many peer-reviewed journal publications support the molecular interaction between D-mannose and FimH-a cellular constituent of the fimbria of E. coli present within the user's body resulting in the prevention of the adhesion to and colonization of the urinary tract by E. coli. In the light of the above rationale, D-mannose intended to prevent UTI does not act by a mechanical or physical action, but by a pharmacological mode of action. Consequently, D-mannose intended to prevent UTIs does not meet the definition of a medical device and should not be qualified as such.

Solution of 8-MOP in extracorporeal photochemotherapy	The 8-methoxy psoralen (8-MOP) solution is intended to be used with a system of extracorporeal photochemotherapy. The 8-MOP solution is placed on the market separately. The intended use of the solution is to treat erythrodermic cutaneous T-cell lymphoma (CTCL). The solution is applied to a sample of the autologous mononuclear cells (MNC) outside of the body and 8-MOP is activated by UV illumination. The treated cells are subsequently re-injected into the patient. The presence of photo activated 8-MOP results in cellular modifications, including cell membrane damage and antigen modification, irreversible cross-linking between 8-MOP and DNA, as well as the formation of covalent bonds with several cytosolic proteins and fatty acids, leading to apoptosis of the treated cells. The principal intended action of this technique is achieved through the activated substance 8-MOP by pharmacological, immunological, and metabolic means. 8-MOP would not be an accessory to any of the medical devices used during the therapy, i.e., the apheresis machine or the UV illumination device, as it is not required for these devices to fulfill their intended function.	In the described case, the 8-MOP solution is intended to treat a specific pathology of the blood (CTCL) by pharmacological, immunological, and metabolic action of 8-MOP on autologous MNC. Consequently, the 8-MOP solution intended to be used with a system of extracorporeal photochemotherapy in the treatment of erythrodermic cutaneous T-cell lymphoma (CTCL), does not fulfill the definition of a medical device and should therefore not be considered as medical device.
Bone void fillers containing animal growth factors	A bone void filler (BVF) made of a resorbable matrix (β-tricalcium phosphate granules (β-TCP, 46 w-%), polyethylene glycol (PEG, 20 w-%), glycerol (31 w-%), stearic acid (1.8 w-%)) and animal growth factors is indicated for the filling of non-load bearing osseous defects of the extremities and pelvis. The manufacturer claims (i) that the matrix provides an osteoconductive scaffold in the bone void for new bone formation and (ii) that the animal growth factors stimulate bone growth and enhance the resorption of the matrix (osteoinduction). It has been categorized by its manufacturer as a medical device and attributed to Class III according to Rule 17 of Annex IX MDD concerning devices manufactured with non-viable products derived from animal tissue. Regulation EU 722/2012 lays down particular requirements in relation to the placing on the market of such medical devices. The TSE risk has thus been assessed according to EN ISO 22442. The pharmacological properties of the growth factors have, however, been ignored in the assessment procedure in spite of the claimed osteoinductive action of the animal growth factors.	With the assumption that the principal intended action of the resorbable BVF is achieved by the osteoconductive matrix and is only assisted in its function by the osteoinductive action of the animal growth factors, the BVF should be qualified as a medical device and attributed to Class III according to rules 8, 13, and 17. In this case, the assessment procedure includes not only: • the risk management measures in connection with tissues of animal origin (as referred to in Directive 2003/32/EC) applied to reduce the risk of infection, but also • the consultation procedure for medicinal products (see Section 7.4 of Annex I of MDD, and part C of MEDDEV 2. 1/3 rev 3).

		If it cannot be clearly established that the osteoinductive nature of the animal growth factors is merely ancillary to the osteoconductive matrix, the BVF will fall within the definition of a medicinal product (see MEDDEV 2.1/3 rev 3 Chapter B.4.1 and MEDDEV 2.1/3 rev 3 Chapter B.2.1) and should therefore not be qualified as a medical device.
Weight management products	Following dietary intake of carbohydrates, the digestive enzyme alpha-amylase is released. It catalyzes the hydrolysis of the alpha-1,4-glucosidic linkages of the starch molecules and eventually converts them into monosaccharides such as glucose which can be absorbed from the intestine. Any excess carbohydrates will be converted into fat following several metabolic routes (e.g., lipogenesis) and stored. Excessive carbohydrate intake over time may consequently result in weight gain and possibly obesity. The extract from Phaseolus vulgaris contains the active ingredient phaseolamin, a proteinaceous substance (glycoprotein) that has been claimed to reduce the activity of alpha-amylase in digesting dietary carbohydrates. When the product is ingested with a starch-containing meal, the active ingredient mimics the molecular structure of starch, and competes with starch molecules for binding to alpha-amylase. Consequently, fewer starch molecules will be processed. The reduced amount of carbohydrate being absorbed by the body will decrease the amount of carbohydrates available for conversion into stored fat. A product based on extract from Phaseolus vulgaris is marketed for general weight management and prevention and treatment of obesity. The mode of action is inhibition of the activity of alpha-amylase and as a consequence, facilitation of weight loss or reduction of weight gain.	The product achieves its principal intended action in the human body by inhibiting an endogenous enzyme, which is an alteration of a metabolic process. Therefore, the product does not meet the definition of a medical device and should not be qualified as such.

10.13. BORDERLINE MEDICAL DEVICE-BIOCIDES

General disinfectants fall under the Biocides Directive. Article 1(2) of this Directive excludes products that are defined or within the scope of Directive 93/42/EEC on medical devices (MDD) and Directive 90/385/EEC on active implantable medical devices (AIMDD). Manufacturers should note that a guidance document exists for borderline cases between the Biocides Directive and the Cosmetics Regulation, as well as a manual of decisions for implementation of the Biocides Directive. That is not discussed as it not within the scope of this volume.

Borderline Device	Background	Outcome
Hand disinfectants	A manufacturer directed a request to the Commission services on the qualification of a range of products. These products are hand disinfectants.	Hand disinfectants do not appear to be qualified as an accessory to a medical device. These products are for disinfecting the hands and not devices. Such products are likely to be covered by other Community legislation, for example the Biocides Directive.
Insect repellent	The product is an insect repellent packaging in a spray and consisting of lactic acid. According to the manufacturer's claims, the product is intended to be applied on the human skin and is intended to prevent diseases by preventing human exposure to diseases transferred by mosquitoes. The manufacturer also claims that the product alleviates injuries on the skin since the product reduces the amount of skin penetration caused by the bites of mosquitoes. According to the manufacturer, the mode of action is that the product acts as a barrier on the skin against mosquitoes and other blood sucking insects, with an effect during four hours. Therefore, the manufacturer intends to put its product on the market as a medical device.	This product is primarily intended to repel insects in order to prevent mosquitoes and insects' bites. This product is not intended to be used principally for a medical use. According to Annex V of Directive 98/8/EC, repellents and attractants products, including those that are used for human or veterinary hygiene either directly or indirectly, are considered biocides products. In addition, insect repellents are considered to have a primary effect on the insects and not on the human body. On the basis of the above, this insect repellent cannot be consider as a medical device.

Multipurpose disinfectants	Disinfectants cover a wide area of uses and, while some are specifically intended for the disinfection of medical devices, others are of a multipurpose use covering the disinfection of various surfaces including floors, walls, sanitary facilities, and sometimes also medical devices. While usually disinfectant products are regulated within the biocides legal framework, those that are specifically intended for disinfecting medical devices fall within the scope of the Directive 93/42/EEC. "Products with a multiple purpose which may be used occasionally in a medical environment are normally not medical devices" (MEDDEV 2. 1/1 paragraph 1.1). "Examples of accessories of medical devices • Disinfectants specifically intended for use with medical devices (e.g., endoscopes). *Note*: Multipurpose disinfectants or sterilization agents are not covered by MDD; they are covered by the directive on biocides." (MEDDEV 2. 1/3 rev 3, paragraph A.2.1.4)	General disinfectants fall under the Directive 98/8/EC on the placing of Biocidal products on the market. This directive will be repealed and replaced by the Regulation (EU) No 528/2012 applicable 1 September 2013.
Brushes and sponges for washing/cleaning nails, hands, and/or harms in hospitals (prior to surgery)	Single use sterile brushes and sponges with or without disinfectants are used by healthcare professionals for washing and cleaning the nails, hands, and/or arms before surgical procedures. Manufacturers claim that, as these products act to prevent infections in patients undergoing surgical procedures, they are medical devices.	Brushes and sponges, with or without disinfectants, for washing and cleaning the nails, hands, and/or arms, used by healthcare professionals, do not qualify as medical devices as they do not meet the definition of a medical device according to Directive 93/42/EEC. Additionally, it should be noted that only products intended specifically to be used for disinfecting medical devices come into the scope of the MDD.

10.14. BORDERLINE MEDICAL DEVICE-COSMETIC PRODUCTS

Borderline Device	Background	Outcome
Tooth whitening or bleaching products	Dental cosmetic products are intended to clean the teeth and/or to whiten/bleach discolored teeth in order to remove the plaque and other residues and/or remove discoloration of the teeth. There may be various claims, such as prevention of odor from the oral cavity or even of some kind of dental caries. Dental cosmetic products may be in the form of solutions, pastes or other forms and are typically intended for use by individuals at home, but also for use in a professional medical environment, e.g., by dentists. The question is whether tooth whitening/bleaching products may be qualified as medical devices, or whether they are simply intended for esthetic purposes/toiletry purpose and therefore cannot be considered as medical devices.	In some cases, in addition to other contributory factors, discoloration of teeth may be caused by a disease. Nevertheless, discoloration of teeth is not considered to be a disease in itself. Besides, application of tooth-whitening/bleaching products is not intended to treat the underlying disease; it only may mask a sign of an underlying disease. Art. 1 of the Directive 93/42/EEC defines that a medical device is intended to be "used for human beings for the purpose of: diagnosis, prevention, monitoring, treatment or alleviation of disease [.]." This definition clearly establishes a link between prevention or treatment and disease. Describing the color of the teeth after treatment does not give an indication about the effectiveness to treat an underlying disease or to prevent a disease. It should be mentioned that, according to the definition of a cosmetic product as laid down in Article 2 of Regulation (EC) No. 1223/2009 on cosmetics products ('Cosmetics Regulation'), tooth whitening or bleaching products qualify as cosmetic products. This has been confirmed in Annex III of this Regulation (entry 12 (d) and (e)), which sets maximum concentrations of hydrogen peroxide, present or released, use conditions and labeling warnings. Therefore, tooth whitening/bleaching products, whether placed within the tooth cavity or placed on the surface of the teeth should not be qualified as medical devices.

Alum styptic pencils	Styptic or hemostatic pencils are sticks made of powdered crystal from an alum block — aluminum sulfate or potassium aluminum sulfate — and a waxy binder, which are pressed into a pencil shape. They are generally used to seal cuts and minor abrasions, especially from razors. They are applied directly to the bleeding site and will stop the bleeding rapidly. The ingredients have astringent properties and act as vasoconstrictor in order to disable blood flow. Styptic pencils are presently (2015) put on the European market both as cosmetic products and medical devices.	Styptic pencils are not considered as cosmetic products, since they are intended to come into contact with injured skin. The aluminum sulfate salts provide a local astringent effect. The action of these salts is based on (1) the precipitation of proteins on the superficial layer of the skin which in turn (2) constricts the tissues, thereby (3) controlling the bleeding. The precipitation is a simple chemical reaction leading to the physical blockage of the damaged blood vessels. As the mode of action of aluminum sulfate salts is other than pharmacological, immunological, or metabolic, styptic pencils should fall under the medical devices regulatory route, namely as Class IIa medical devices according to Rule 4, third indent of Annex IX of Directive 93/42/EEC.

10.15. ACCESSORY TO A MEDICAL DEVICE OR AN *IN-VITRO* DIAGNOSTIC MEDICAL DEVICE

Let us recap the definitions according to the Regulations:

Article 1 MDD (b) 'accessory' means an article which whilst not being a device is intended specifically by its manufacturer to be used together with a device to enable it to be used in accordance with the use of the device intended by the manufacturer of the device.

Article 1(2)c IVDD "'accessory' means an article which, whilst not being an in vitro diagnostic medical device, is intended specifically by its manufacturer to be used together with a device to enable that device to be used in accordance with its intended purpose. For the purposes of this definition, invasive sampling devices or those which are directly applied to the human body for the purpose of obtaining a specimen within the meaning of Directive 93/42/EEC shall not be considered to be accessories to in vitro diagnostic medical devices."

Borderline Device	Background	Outcome
Hemodialysis water test strips	The matter at hand concerned the question whether the following two products can be considered as an accessory to a hemodialysis machine which is a medical device. Both products are used together with the hemodialysis machine, one is a residual peroxide reagent strip, which confirms, when used, that the residue of disinfection agents used in the hemodialysis machine has been reduced to safe levels. The other is a reagent strip used to test the water in the hemodialysis machine to ensure that the level of water hardness has been reduced to a level where it is safe to proceed with hemodialysis.	From the information provided by the manufacturer of the strips, they do not enable the hemodialysis machine to be used. The strips are used for testing and are not necessary for the functioning of the machine. According to Article 1(2) b of Directive 93/42/EEC and the developed guidance (MEDDEV 2.1/1) the decisive criterion to decide whether a product is an accessory to a medical device is whether or not the product is specifically used together with a medical device to enable it to be used in accordance with the use intended by its manufacturer. The notion "enabled it to be used" implies that the accessory is necessary for the medical device to function. Therefore, such strips are not considered to be 'accessories' of a medical device within the meaning of Article 1(2) (b) MDD. If a manufacturer can claim, substantiated with a solid reasoning that the strips are necessary for the proper functioning of the machine, then these products might be qualified as 'accessories.'

Surgical instrument decontamination products	A manufacturer directed a request to the Commission services on the qualification of a range of products. The products in question are surgical instrument decontamination products.	Surgical instrument decontamination products are covered by the definition of accessories to medical devices in Article 1(2) b of Directive 93/42/EEC. The respective guidance (MEDDEV 2. 1/1) expressively mentions disinfectants specifically intended for invasive medical devices as one example for accessories. For the classification of these products see paragraph 8.9
Dental water line disinfectants	A manufacturer directed a request to the Commission services on the qualification of a range of products. The products in question are Dental Water Line Disinfectants.	Dental Water Line Disinfectants are covered by the definition of accessories to medical devices in Article 1(2) b of Directive 93/42/EEC. For the classification of these products see paragraph 8.10
Sterilization indicators	The sterilization procedure is monitored routinely by using chemical and biological indicators to evaluate the sterilizing conditions and indirectly the microbiologic status of the processed items.	Sterilization indicators monitor the performance of the sterilizer. They do not affect the sterilization procedure and only provide additional information to the user. Sterilization indicators do not fulfill either the definition of a medical device laid down in Article 1(2)a of Directive 93/42/EEC or the definition of an accessory laid down in Article 1(2) b of Directive 93/42/EEC as they are not intended specifically to be used together with a device to enable it to be used in accordance with its stated use.

Microplate washers	The product is a microplate washer intended to wash standard flat-bottom 96-well plates or microstrips during the enzyme-linked immunosorbent assay (ELISA) in diagnostic laboratories. This microplate washer is a fully programmable device ensuring multi-step solution ripening and aspiration. The manufacturer claims that this microplate washer falls within the scope of the Directive 98/79/EC and meets the definition of *in vitro* diagnostic medical device because this microplate washer is intended for the performance of the ELISA protocols and therefore specifically intended for the use for *in vit*ro diagnostic examination. However, this microplate washer is not used for direct examination of specimens and calculation of the ELISA results and no specific IVD requires this washer to be used to enable the IVD to perform its intended purpose.	This microplate washer is not intended to be used *in vitro* for the examination of specimens, solely or principally for the purpose of providing information concerning a physiological or pathological state or concerning a congenital abnormality or to determine the safety and compatibility with potential recipients or to monitor therapeutic measures. No specific ELISA requires this washer to be used for *in vitro* diagnostic tests. In addition, MEDDEV 2.14/1 rev.1 states in this context: "if, however, the product does not in fact possess specific characteristics that make it suitable for one or more identified *in vitro* diagnostic examination procedures, then the manufacturer is not free to bring it within the scope of the IVDD merely by affixing the CE marking to it. In other words, a manufacturer is not able to bring within the scope of the IVDD a product that, in reality, is a piece of general laboratory equipment simply by affixing the CE mark to it." If this microplate washer possessed such specific characteristics, the manufacturer would have to demonstrate them and the link with one or more specifically identified *in vit*ro diagnostic tests. On the basis of the above, this microplate washer should be regarded as a general laboratory product. This is excluded from the IVDD by Article 1(2) b IVDD.

Automated external defibrillator storage units	Automated external defibrillators (AEDs) are intended by their manufacturer for use in emergency situations and are usually designed to function as intended within certain environmental conditions, e.g., within temperature and humidity ranges that are specified in the instructions for use and on the device labeling. To ensure AEDs are available for use in an emergency situation, they are often stored in public locations (e.g., shopping centers and public streets). Due to the varying environmental conditions that the AED can be exposed to in these locations, AEDs are stored in storage units that claim to maintain the AEDs within the recommended environmental conditions. An AED is a medical device in accordance with the MDD. The question is whether the AED storage units could be considered as an accessory to the AED.	An AED storage unit is intended to be used with an individual AED to store it in the environmental conditions in which the AED was designed to function. The AED storage unit protects the AED from extreme environmental conditions in the location where it is stored so that the AED performs as intended when needed in an emergency situation. An AED storage unit that is intended to maintain the specified environmental conditions required for an AED to perform as intended should be qualified as an accessory to a medical device as per Article 1(2) b of the MDD. An AED storage unit that uses a power supply to achieve its intended purpose should be classified as Class I under Rule 12 of the Annex IX of the MDD. If no power supply is used, the unit should be classified as Class I under Rule 1. On the other hand, an AED storage unit that is not intended to maintain the required environmental conditions for the AED should not be qualified as an accessory to a medical device.

10.16. CLASSIFICATION OF BORDERLINE MEDICAL DEVICES

While determining whether a borderline product is a medical device or medicine (or neither) can be confusing, this is not the last step in the process. After classifying a product as a borderline medical device, the next step is a classification of the device itself. Manufacturers should consult with the regulation and the MEDDEV 2.4/1 for more detailed guidance concerning the classification rules. Examples of device classification for some borderline medical devices are presented below as taken from the Borderline Manual.

Borderline Device	Background	Outcome of Classification
Lightbox indicated to treat the seasonal affective disorder (S.A.D)	The product in question is a lightbox that emits bright light, and the manufacturer states that light therapy is 'a convenient and effective way of compensating for the lack of light without resorting to medication.' The manufacturer also states: 'in autumn and winter, the seasons with the least sunlight because the days are shorter, increased symptoms resulting from light deprivation may be experienced. Even standard artificial lighting in buildings cannot compensate for a shortage of natural light. These statements are effectively claiming for treatment of seasonal affective disorder (S.A.D.), which is a recognized medical condition, and therefore this product is considered a medical device.	In the classification of this product, it must be decided whether it performs any active action as defined in Annex IX I-Definition 1.5 'Active therapeutic device.' That is, does it support, modify, replace, or restore biological functions or structures with a view to treatment or alleviation of a disease? If the device performs an 'active' function, then it is Class IIa.
Oxygen delivery	These products are intended to deliver oxygen to the patient and are connected to active devices such as ventilators, anesthetic machines, etc., and/or to a pressure regulator. These devices may be connected to the different kind of oxygen delivery systems with regulators; via oxygen piping to the oxygen supply/oxygen delivery center, to oxygen bottles/oxygen cylinders or to oxygen concentrators.	In the classification of this product, Class IIa is appropriate based upon Rule 2, 5, or 11 of Annex IX.

Examination gloves coated with polyhexamethylene biguanide (PHMB)	The case relates to examination gloves with PHMB which is a broad-spectrum bactericide. This substance is also used as an ingredient in various products (contact lens solutions and surgical scrubs and swimming pools). The intended use is to reduce bacterial transfer between the healthcare professional and the patient. The gloves would be single-use.	Examination gloves are usually considered to be Class I medical devices, however MEDDEV 2.1/3 in Section A.5 states that 'wound dressings, surgical or barrier drapes (including tulle dressings) with antimicrobial agent' are considered to be devices incorporating medicinal substances and therefore Class III devices. Antimicrobial agents on surgical or barrier drapes intended to come in to contact with the patient have no 'ancillary' effect on the patient and neither would an antimicrobial coating on an examination glove, however the MEDDEV implies that these examination gloves with a PHMB coating should be considered as Class III medical devices. Medical devices may incorporate substances as an integral part which, if used separately, may be considered to be a medicinal product. This is specifically addressed in Article 1(4) MDD which makes it clear that such products are devices, provided that the action of the medicinal substance is ancillary to that of the device, as reflected in the product claim
		and as supported by the scientific data provided by the manufacturer of the devices. Rule 13 places these devices in Class III. In essence, two issues need to be considered: a) is the substance (PHMB), if used separately a medicinal product; b) is the substance liable to act on the human body with action ancillary to that of the devices? a) Taking into account the published literature, it can be concluded that the PHMB is a substance which could be administered topically to human beings in view to restore or modify physiological functions by mainly means of pharmacological action (e.g., treatment of Acanthamoeba keratitis). As such it could be regarded as a medicinal product in accordance with Article 1(2) of Directive 2001/83/EC as amended. b) The risk that the PHMB acts on the patient highly depends on the intended use of these gloves. For example, an examination of a wound or a mucous membrane will lead to a considerably increased risk of action of PHMB on the patient. On the basis of the above and taking into account the Rule 13, the classification of these gloves as Class III would appear the most appropriate.

Picture archiving and communication systems (PACS)	Basically, a PACS workstation is specifically designed to be networked with a wide variety of diagnostic imaging systems, e.g., x-ray, nuclear medicine, magnetic resonance imaging (MRI) or ultrasound, as well as laboratory or hospital information systems. It does not contain controls for the direct operation of a diagnostic imaging system and is designed to receive, archive, and transmit data both online and off-line. It is typically located at a site remote from imaging systems and is configured to provide limited or extensive capabilities to further process, manipulate, and/or view patient images and information collected from diagnostic imaging systems. The manufacturer of the PACS states that the system does not influence the radiation of the diagnostic x-ray machine. Generally speaking, there are various types of PACS: (a) PACS used for viewing, archiving, and transmitting images. (b) Where the post-processing of the image for diagnostic purposes is such as: • image processing functions which alter the image data (e.g., filtering, multiplanar reconstruction, 3D reconstruction) • complex quantitative functions (e.g., arterial stenosis evaluation, ventricular volume calculation, calcium scoring, automatic indication (detection) of potential lesions (c) With image-enhancing by controlling image acquisition	In cases where the PACS falls under the definition of a medical device, i.e., is specifically intended by the manufacturer to be used for one or more of the medical purposes set out in the medical device definition, the following situations can be foreseen: (i) In relation to PACS (a) intended by its manufacturer to be used for viewing, archiving, and transmitting images, it is considered that applying Rule 12 could be appropriate, and accordingly, this type of PACS is generally classified as Class I medical devices. However, PACS that are only intended for archiving or storage of data may not fall within the definition of a medical device provided that data is not manipulated. (ii) Those types of PACS (b) which drive a device or influence the use of a source device fall automatically in the same class in accordance with implementing Rule 2.3, which classifies them as Class IIa or IIb. If this type of PACS does not drive or influence the use of the source device, this type of PACS can be classified under Rule 10 if such PACS are intended to allow direct diagnosis, classifying them as Class IIa. (iii) PACS with image enhancing by controlling image acquisition (c) should fall into the same class as the source device. This is based upon, firstly, implementing Rule 2.3" Software, which drives a device or influences the use of a device, falls automatically in the same class." and the last paragraph of MEDDEV 2.4/1-rev. 8, Section 3.2 stating that: "Standalone software, e.g., software which is used for image enhancement is regarded as driving or influencing the use of a device and so falls automatically into the same class. Other standalone software, which is not regarded as driving or influencing the use of a device, is classified in its own right." Applying this classification rule and the interpretation of the MEDDEV allows this type of PACS to be classified as Class IIa or IIb medical devices according to the classification of the device itself.
Blood refrigerators, freezers, and defrosters	The product in question concerns blood product cooling devices/blood bank refrigerators. A blood storage refrigerator and a blood storage freezer are devices intended for medical purposes that are used to preserve blood and blood products by storing them at cold or freezing temperatures. Plasma defrosters are designed for defrosting of blood plasma.	Blood storage refrigerators, freezers, and defrosters sold for the specific intended purpose of dealing with blood should be medical devices in their own right. The cooling device or refrigerator store substances that will be eventually delivered into the body and are Class IIa (Rule 2). The plasma defrosters are in Class I (Rule 1).

Warming blankets	This matter relates to the classification of warming blankets. It concerns a manufacturer who markets warming blankets on the basis of warm air, produced by an extra warming device. The manufacturer considers the blanket to be an accessory to the active device and as such the manufacturer considered it Class I.	1: Medical device or not? The medical purpose has to be clearly identified and substantiated to qualify these products as medical devices 2: If the blanket and the generator are sold as a single medical device: Classification Rule 9 would classify these products as class IIa or Class IIb medical devices depending on the state of the patient (which is inherently linked to different levels of risks). When the product is intended to be used on an unconscious patient (who therefore cannot remove the blanket), e.g., in reanimation services, the device is to be classified as a Class IIb medical device. If the device is intended to be used on a conscious patient (who therefore can react), the device is to be classified as a Class IIa medical device. 3: If the blanket and the generator are sold separately: The blanket sold separately cannot be considered as an active medical device and is a Class I medical device in accordance with Rule 1. The generator sold alone would be Class IIa or Class IIb in accordance with classification Rule 9. The manufacturer will have to specify the use of such a generator.
Products evaluating the condition of respiratory muscles	The product assists in evaluating the condition of the respiratory muscles. It is a small battery-operated device that performs two tests: the Pmax and the Sniff tests. Both tests measure pressure. The Pmax test operates through the mouth through a voluntary respiratory maneuver. This test requires a maximal respiratory effort by the patient. The test starts with a deep breath filling the full capacity of the lungs, and then the patient tries to breathe out into a plastic tube which is closed from all sides except the connection to the mouth. The device measures the maximum pressure that is developed in the tube and sustained for at least one second. This pressure is then compared to normal pressures. This is done for both expiratory and inspiratory pressures. The Sniff test measures the pressure that develops in the nose during a sniff maneuver. Both tests are supposed to give similar results.	The product measures how fast a person can exhale air. It is one of many tests that measure the function of the airways, which are commonly affected by diseases such as asthma. This product is intended to measure the condition of the respiratory muscles and as such is to be classified as a Class IIa medical device in accordance with classification Rule 10 third indent.

	It gives an indication whether the respiratory muscles are working properly. Bad readings do not necessarily indicate a disease but an indication that further clinical investigation might be appropriate. Neither does it monitor any physiological process since it does not allow airflow through it, so it cannot monitor respiration.	
Neutral electrodes for high frequency surgery	The issue relates to the classification of neutral electrodes which are accessories for high frequency (HF) surgery. In general, there are two types of electrodes used in high frequency surgery: those which are 'active' concentrate the energy and convert it into heat and those which are 'neutral' and simply transfer the energy between two points. The MEDDEV does not make this distinction and places both electrodes active and neutral under Rule 9, consequently they are both considered to be in Class IIb. This is the result from discussions a few years ago, and that is why the MEDDEV puts both electrodes in the same Class IIb. However, manufacturers do make this distinction.	Current guidance (MEDDEV 2.4/1 under Rule 9) indicates that all such electrodes should be considered as Class IIb products, irrespective of their nature (active or neutral) as they may be potentially hazardous. It is concluded that the neutral electrodes are medical devices which involve potentially hazardous exchange of energy and should be Class IIb medical devices.
Surgical instrument decontamination products	A manufacturer directed a request to the Commission services on the classification of a range of products. The products in question are surgical instrument decontamination products.	These products should be classified according to Rule 15 of Annex IX of Directive 93/42/EEC. This rule has been further developed in MEDDEV 2.4/1, according to which this rule covers substances used principally in a medical environment to disinfect medical devices. Examples listed in the MEDDEV 2.4/1 include disinfectants specifically intended for instance for endoscopes or hemodialysis equipment, sterilizers specifically intended to sterilize medical devices in a medical environment, and washer-disinfectors.
Dental water line disinfectants	A manufacturer directed a request to the Commission services on the qualification of a range of products. The products in question are Dental Water Line Disinfectants.	These products should be classified according to Rule 15 of Annex IX of Directive 93/42/EEC. This rule has been further developed in MEDDEV 2.4/1, according to which this rule covers substances used principally in a medical environment to disinfect medical devices.

Dental curing lights	The dental curing lights are intended for curing of dental filling substances in situ. A number of dental filling materials need for hardening (a kind of polymerization) after application to the tooth to be treated with light. During this application of light, the energy transmitted with this light is absorbed by the filling material as well by the surrounding parts of the body (surface of the tooth, other neighbored fillings and crowns, internal part of the tooth surrounding the filling which is warming up, gum if the filling is close to the gum). It is not possible to avoid the surroundings of the filling to be treated together with the filling; this is an undesired but unavoidable and accepted side effect. Because of the considerable changes in the design of the lights, it is questioned whether a reclassification would be needed.	It is confirmed that no reclassification is needed and that these products shall be considered as Class I medical devices in accordance with classification Rule 12. Also, the guidance MEDDEV 2.4/1 on classification lists the example of "dental curing light" under classification Rule 12.
Bacterial/viral filter for use on patient undergoing pulmonary function testing	The product is a bacterial/viral filter indicated for single use on patients undergoing pulmonary function testing, and its main function is to avoid secretion and moisture deposition in the equipment, reducing the need for frequent decontamination procedures. The equipment is a spirometer, a Class IIa active medical device, according to Directive 93/42/EC. The manufacturer states that the product should be classified as Class I, because the product does not interfere at all with the diagnostic function of the test machine, and because the energy used by the active device (spirometer) is not exchanged with the patient. In consequence, the manufacturer applies the rule for devices not intended for connection to an active medical device.	The product is an invasive device since it is partially inserted inside a natural body orifice (patient's mouth). The product is intended for transient use, as the duration of contact seems to be usually less than 60 minutes and is intended for connection to a Class IIa active device (spirometer). On the basis of the above and taking classification Rule 5 into account, this device should be classified as a Class IIa medical device.

Hydrocolloid plaster for blisters	Blisters are resulting from frictional forces that mechanically separate epidermal cells and dermal cells and the separation fills with a fluid. The epidermal cells will be weakened. The blister management could be done by protecting the blister without puncture it or puncturing the blister with a sterile needle and removing the skin on the blister area before placing the plaster. Hydrocolloids are more effective than ordinary first aid plasters and claim that they promote rapid healing. Indeed, some of hydrocolloid dressing mentions the principle known as moist wound healing to provide the ideal treatment for blisters. Moisture enhances wound re-epithelization since occlusive dressings that maintain moisture and warmth were optimal for healing.	A blister is injured skin. Hydrocolloid acts by its proprieties on the microenvironment of the blister and is not placed on a blister to act simply as a mechanical barrier for absorption of exudates. On the basis of the above and taking classification Rule 4 and guidance MEDDEV 2.4/1-rev 8, part 2 into account, it is considered that hydrocolloid which claims promoting rapid healing should be classified as Class IIa medical devices.
Movement monitor for babies	The manufacturer claims that this product is recommended for all babies, especially during their most vulnerable first 6 months, to help guard against life-threatening events such as Apnea and Sudden Infant Death Syndrome (SIDS or Cot Death). A little vibrating motor is used to stimulate babies in neonatal wards that suffer apnea episodes and can indicate if they are rhythmic or general movements. If the baby becomes dangerously inactive, this product will provide a small tactile stimulation even before human intervention. If breathing effort stops, slows down too much, or becomes too shallow, the built-in stimulator will gently stir the baby to breathe, failing which a loud alarm will alert the nearest adult and consequently should be help by a doctor. The problem is to know whether this product could be considered as a general product or as a medical device or as an active medical device.	This product is intended to help preventing life-threatening events such as apnea and sudden infant death syndrome (SIDS or Cot Death). Taking this medical purpose into account, this product fulfills the definition of a medical device. This product is to be connected with a source of electrical energy and could then be considered as an active medical device. According to classification Rule 10, this product should be classified as a Class IIb medical device.

Medical devices containing silver	There are many medical devices being placed on the market containing silver (e.g., wound dressing containing silver, coated catheters, or secondary bandage to cover primary dressings…). The claims for these products are usually that the silver is acting as an antimicrobial agent in an ancillary manner. Silver is well known as an antimicrobial agent and is active at low levels. It is considered to be a medicinal substance in its own right when such claims are made. The question is whether or not silver containing medical devices should be classified as Class III medical devices under classification Rule 13 of Annex IX.	In general, medical devices containing or coated with silver, specifically those making antimicrobial claims, should be considered as Class III medical devices under classification Rule 13 of Annex IX. This is because in order for the silver to act as an antimicrobial agent, it has to be in an available form and therefore would be liable to act on the body. The Directive 93/42/EEC states that if a device incorporates as an integral part a medicinal substance and that this substance is liable to act on the body with ancillary action to that of the device, then the device should be classified as Class III medical device. The Directive does not state that the medicinal substance must be intended to act on the body. The fact that the silver is not intended to reach the body or act on the body does not preclude the silver from being liable to act on the body. There may be some exceptions to this: If a device does not come into direct contact with the body (for example a secondary bandage intended to cover primary dressings) then a lower classification might be possible, provided that the manufacturer has clear data to support the claims being made and the fact that the silver is not liable to act upon the human body. If it is claimed that the silver in the product will not act on the body or that it is intended to maintain specific characteristics of the device and would therefore not have any ancillary action, the manufacturer must demonstrate the claim via clinical and scientific data, using suitably rigorous testing to prove that the silver does not leach out of the device. If there is clear data to support such a claim, then a product could potentially be classified in a lower class. In the absence of valid data to support such a claim, then the product would remain Class III under Rule 13.

Ethyl chloride spray for local refrigeration anesthesia	A manufacturer of a spray for local anesthesia containing ethyl chloride wants to place this product on the market as a medical device. Several other products with a similar intended use and mode of action are already placed on the market as medical devices but with some discrepancies regarding their classification. In some countries, ethyl chloride is regulated as a medicinal product because of its toxicity and its narcotic properties. Ethyl chloride is used as a local anesthetic in minor operative procedures and is used also to alleviate pain associated with bruises, contusions, etc.… The rapid vaporization of ethyl chloride when applied as a spray to the skin produces freezing of superficial tissues to –20°C, which results in insensitivity of peripheral nerve endings and a local anesthesia.	In the present case, the principal mode of action of ethyl chloride is not pharmacological, immunological, or metabolic, and therefore this product could be qualified as a medical device. As stated in MEDDEV 2.4/1 rev.9, medical devices using pre-stored gases and/or vacuum as a power source are regarded as active devices. Consequently, this product could be qualified as an active medical device and, according to Rule 9 of Annex IX of Directive 93/42/EEC, could be classified as a Class IIa medical device.
Pathogen inactivation system for platelets	These specific systems for storing platelets are used to protect patients against transfusion-transmitted disease by inactivating pathogens in platelets. The system is designed to inactivate viruses, bacteria, other pathogens, and white blood cells in platelets intended for transfusion. This system is composed of a blood bag, a chemical compound and, if necessary, an illumination device. The compounds have a high affinity for nucleic acids of viruses, bacteria, and others pathogens and prevent the replication of the pathogens by damaging the nucleic acid. The chemical compounds could be psoralen-derivatives, riboflavin, ethylene imines and methylene blue. Psoralen-derivatives and riboflavin require a photo-activation by an illumination device to be active. After the pathogen inactivation process, the chemical compounds could be removed, and the platelets are transferred to the final blood bag where they are stored until transfusion. The question is on the classification of such devices, the blood bag with chemical compounds and the illumination device.	According to MEDDEV 2.1/3 rev 3, "systems intended for the collection, storage, and preservation of blood or blood components and as an ancillary function, the treatment of blood or blood components where this effect is achieved outside the human body, are classified as devices provided that any residual material is not intended to achieve its intended effect when the blood or cells are reintroduced into the body, e.g., systems incorporating chemicals activated by light to reduce the viral load where the quantity of chemical remaining has no intended effect when transfused." According to MEDDEV 2.4/1 rev 9, blood bags are Class IIb except if they "have a function greater than for storing purposes and include systems for preservation other than anti-coagulants the other rules (e.g., Rule 13) may apply." Therefore, if the compound has an ancillary pharmacological or metabolic action, the blood bag with the compound has to be classified as a Class III medical device under Rule 13. In the present case, the compound which inactivates pathogens is used to provide an antimicrobial effect and not for the purpose of the preservation of the solution. This antimicrobial effect is considered to be a pharmacological action, and therefore the blood bag with the compound should be classed Class III under Rule 13.

		The illumination device is considered to be an accessory of the device as it enables the action of the blood bag with the chemical compound and therefore could be classified as Class IIa medical device under Rule 3, as the illumination process modifies the biological composition of blood and consists on exchange of energy.
Pre-transfusion confirmatory tests	Pre-transfusion devices are made of a support on which anti-A and anti-B reagents are applied. Those devices are used by nurses and doctors for the pre-transfusion compatibility testing at the patient's bedside. The purpose is to verify the ABO system compatibility between the recipient and the red blood cell component to be transfused immediately before transfusion to prevent any incidental incompatibility. The simultaneous comparison of agglutination between the recipient's blood and the blood of the donor's bag allows a last verification of the ABO system compatibility between recipient and donor. These devices are considered as *in vitro* diagnostic medical devices. The question is whether these devices are falling under Annex II list A of Directive 98/79/EC, which includes "Reagents and reagent products, including related calibrators and control materials, for determining the following blood groups: ABO system…."	Although the Pre-transfusion devices are not specifically used for the determination of the ABO system blood groups, those devices include anti-A and anti-B reagents used for the determination of the ABO system blood groups, which are falling under Annex II list A. Therefore, these devices are falling under Annex II list A of Directive 98/79/EC.

Eye drops regulated as medical devices	The ocular surface is a physiological complex fundamental in the maintenance of a good visual function, so its integrity must be properly safeguarded irrespective of the use of contact lenses. The use of substances that increase the retention time of the lubricants, tend to increase the contact time between the lubricant and the ocular surface. According to point 3.1.2.2 of MEDDEV 2.4/1 rev.9 on the Concept of continuous use, if it cannot be demonstrated that components of the device are totally eliminated in the interval between uses, this is also considered as an immediate replacement. The continued use in situations where there is no lasting relief of symptoms as well as the continuation of aggressive conditions (environmental or not/or due to a number of medical conditions) can lead to the permanent and/or high frequency use (with products whose rate of elimination is unknown), setting the conditions for the continued use of these lubricants. Considering eye lubricants as examples of medical devices (in which there is no demonstration of their total elimination between applications) destined for continuous use.	According to Rule 15 of Annex IX of Directive 93/42/EEC, all devices intended specifically to be used for disinfecting, cleaning, rinsing or, when appropriate, hydrating contact lenses are in Class IIb, and in MEDDEV 2.4/1 rev.9, comfort solutions are given as an example to Rule 15. When qualified as medical devices, eye drops intended for the lubrication of the eyes, should be classified on a case-by-case basis, and would be Class IIa or Class IIb, in accordance with classification Rule 5 of Annex IX of Directive 93/42/EEC, depending on if they are intended for short or long-term use.
Wound irrigation solutions containing antimicrobial agents	Irrigation solutions intended for mechanical rinsing are regarded as being medical devices. Topical disinfectants for use on humans are not considered to be medical devices but rather medicinal products. A number of irrigation solutions intended for mechanical rinsing contain ingredients such as chlorhexidine, cetrimide, iodine, hypochlorous acid (HOCl), free chlorine (chlorine/chloride ion Cl_2/Cl^-), hydrogen peroxide, hypochlorous acid, hydrogen peroxide, chlorine dioxide, sodium hydroxide, sodium chloride and sodium carbonate. Many of these ingredients have an antimicrobial effect on the body. A question arose with regard to the qualification of these products as medical devices and their classification.	MEDDEV 2.1/3 rev.3 states that irrigation solutions for mechanical rinsing are considered to be medical devices unless the principle intended purpose is to provide a local antimicrobial effect. The majority of wound irrigation solutions with antimicrobial action are intended primarily as wound irrigation solutions to mechanically rinse the wound whilst also reducing the bacterial load. The antimicrobial effect may be considered as an ancillary action. Antimicrobial agents or substances are generally considered to be medicinal substances when intended for a medical purpose. Therefore, a wound irrigation solution containing an antimicrobial agent, irrespective of the amount of the antimicrobial substance liable to act on the body, will be considered to be Class III according to classification Rule 13.

Contact lenses	There are three main types of contact lenses: • Daily disposable (single use) contact lenses: these single-use devices are worn only during the waking period of one day and then disposed of; -Contact lenses for daily wear (planned replacement): these contact lenses are reusable devices. They are worn only during waking periods; taken out overnight for cleaning, disinfected during the sleeping period, and put back in the eye the following day; • Contact lenses for extended or continuous wear: these lenses are worn continuously during successive waking and sleeping periods.	The overnight period when lenses are cleaned and disinfected is considered as a discontinuation of the device use. For the determination of the duration of use, only the specified time period of uninterrupted wear of the lens (e.g., 16 hours) needs to be taken into account. On the basis of the above, contact lenses are classified as follows: • Daily disposable (single use) contact lenses are classified as Class IIa medical devices according to classification Rule 5, 2nd indent; • Contact lenses for daily wear with discontinuation of use overnight are classified as Class IIa medical devices according to classification Rule 5, 2nd indent; -Contact lenses for overnight wear with discontinuation of use during the day are classified as Class IIa medical devices according to classification Rule 5, 2nd indent; • Contact lenses for continuous wear for up to 30 days are classified as Class IIa medical devices according to classification Rule 5, 2nd indent; • Contact lenses for continuous wear for more than 30 days as classified as Class IIb medical devices according to classification Rule 5, 3rd indent.
Paraffin oil for IVF/ART procedure	Paraffin oil is placed over the culture media containing the human embryos during an ART or IVF procedure. The paraffin oil overlay is intended to protect the culture medium from evaporation in the incubator, to reduce gas, temperature, and pH fluctuation in the medium surrounding the embryos. Besides, the paraffin oil eases the examination of embryos during culture under the microscope.	The paraffin oil protects the medium and embryo from environmental conditions by creating a physical barrier like containers for organs or body tissues and is intended to be used for IVF/ART procedure. When intended to be used for IVF/ART procedure, the paraffin oil should be qualified as a medical device. Besides, it has been established that the paraffin oil by its relatively close contact between the oil and the embryo could interfere on the embryo development. Thus, by similarly with containers for organs or body tissues, the paraffin oil should be classified as a Class IIa medical device in accordance with classification Rule 2 second indent.

Dental abutments	Three parts make up an "artificial" tooth: the crown or cap, the dental abutment, and the implant. This entry refers to final dental abutments which are usually called dental implant abutments or prosthetic abutments for dental implant and does not cover healing abutments which are placed during a variable time before a final abutment. Dental abutments are connecting elements between the dental implant and the crown. The implant is inserted directly into the bone of the jaw. The abutment is fixed to the implant and is in contact with the gum, in the surgical cavity. In the final stage of getting the dental implant, the crown is built above the gum around the other part of the abutment. The question arises as to whether dental abutments should be classified as Class IIa or IIb medical devices.	MEDDEV 2.4/1 rev. 9 relating to the Rule 8, first hyphen, of Annex IX to Directive 93/42/EEC states that examples of medical devices to be placed in the teeth such as bridges and crowns, dental filling materials and pins and dental alloys ceramics and polymers are to be classified as Class IIa medical devices. Since they are directly placed in the gum, dental abutments should be considered as implantable devices, as well as dental implants. According to Rule 8 of Annex IX to Directive 93/42/EEC, dental abutments should be classified as Class IIb medical devices.
Autologous platelet preparation system	This product is a system which is used in the preparation of autologous platelet rich plasma (PRP) for injection/application in/on the skin surface or wound and is claimed to promote accelerated healing and rejuvenation. The system comprises several devices such as preparation tubes and filters, syringes, hypodermic needles, and blunt needles. Each of these devices is CE marked. The patient's blood is drawn into the preparation tube which contains a gel and an anti-coagulant solution (ACD). The preparation tube is then centrifuged to separate the patient's blood sample into two phases: red blood cells (erythrocytes) and plasma, which contains the platelets. The gel, which has a specific gravity between that of the plasma and red blood cells, acts as a barrier to prevent the two phases reconstituting. The plasma phase consists of two 'layers': a top layer of Platelet Poor Plasma (PPP) and a bottom layer of Platelet Rich Plasma (PRP). The PPP is removed by inserting a blunt needle syringe into the plasma and then discarded. The remaining PRP is extracted through a filter sleeve, which is placed within	Anti-coagulant solutions (ACD) and preservative solutions for human blood have a European Pharmacopeia monograph of 0209 and would be considered to be a medicinal product as defined in Article 1 of the Directive 2001/83/EC. Considering the anti-coagulant solution (ACD) is mixed with the blood during manipulation of the patient sample, the anti-coagulant solution (ACD) is liable to act on the human body when PRP is introduced in or used on the human body. Therefore, the platelet preparation tubes, containing the anti-coagulant solution (ACD), may be classified as a Class III medical device under Rule 13 in the case where either the PRP infusion that the patient receives contains medicinal substances or in circumstances where the manufacturer cannot appropriately demonstrate that such medicinal substances have been completely removed from the PRP infusion. In the case where this can be demonstrated, the second indent of Rule 3 of Annex IX to Directive 93/42/EEC may apply due to the treatment process consisting of a centrifugation step. Analogy to Rule 18 of the above-mentioned Annex which classifies blood bags, including those containing or coated with an anticoagulant, as Class IIb medical devices was made. This rule, however, is a special derogation rule which is exclusively applied to blood bags and cannot be extended to other product which may contain an anticoagulant solution.

	the preparation tube, using the blunt needle syringe. The blunt needle is removed from the syringe and replaced with a hypodermic needle which is then used to inject the PRP into the patient. The manufacturer has classified and CE marked the preparation tube, containing the gel and anti-coagulant solution (ACD), as a Class IIa medical device under Rule 3 as he means the tube to be a non-invasive device modifying the composition of blood (with the aid of centrifugation) and intended for infusion into the body. Anti-coagulant solution (ACD) is considered to be a medicinal product as defined in Article 1 of Directive 2001/83/EC and according to Rule 13 of Annex IX of Directive 93/42/EEC 'all devices incorporating, as an integral part, a substance which, if used separately, can be considered to be a medicinal product as defined in Article 1 of the Directive 2001/83/EC, and which is liable to act on the human body with action ancillary to that of the devices, are in Class III.' However, as per MEDDEV 2.4/1 Rev. 9, this rule does not cover those devices incorporating substances which under other circumstances may be considered as medicinal substances, but which are incorporated into the device exclusively for the purpose of maintaining certain characteristics of the device and which are not liable to act on the body. The appropriate classification of the preparation tube, containing the gel and anti-coagulant solution (ACD) was queried, specifically the appropriateness of Rules 3 and 13. The appropriate classification of the filter sleeve was also queried.	Sleeve filters should be classified as a Class IIa medical device under Rule 3 of Annex IX of Directive 93/42/EEC according to which all non-invasive devices intended for modifying the biological or chemical composition of blood, other body liquids or other liquids intended for infusion into the body are in Class IIb, unless the treatment consists of filtration, centrifugation or exchange of gas or heat, in which case they are in Class IIa. In the scenario where the system is placed on the market containing one or more medical devices not bearing valid CE marking, the system shall be treated as a device in its own right. In the same manner, where the manufacturer of the system changes or modifies the IFU for the medical devices included in the system, or creates a new IFU in order to claim the production of PRP, the system shall be treated as a device in its own right. The applicable conformity assessment procedure shall be determined by the medical device component with the highest classification, i.e., the platelet preparation tubes containing the anti-coagulant solution (ACD). Therefore, the entire system would be classified as a Class III medical device under Rule 13 where either the PRP infusion that the patient receives contains medicinal substances or in circumstances where the manufacturer cannot appropriately demonstrate that such medicinal substances have been completely removed from the PRP infusion.

Antimicrobial photodynamic therapy (APDT) systems	Photodynamic laser-based disinfection systems, activated by a topical photosensitizer, are intended for the decolonization of potentially pathogenic bacteria, including methicillin-resistant S. aureus (MRSA), from the oral cavity or anterior nasal passages. The topically applied photosensitizer, such as methylene blue or toluidine blue, stains bacteria by binding with microbial cell wall components. Light, at a specifically defined wavelength, is absorbed by the topically applied photosensitizer molecules in the presence of oxygen. This causes the photosensitizer molecules to undergo excitation and electronic state transitions, converting the sensitizer to its photoactive triple state. The excited photosensitizer immediately transfers energy to surrounding molecular oxygen, thereby producing reactive oxygen species (ROS) which are responsible for lethally disrupting the microbial cell wall. These ROS products are very short-lived, and the ROS-production process ceases immediately upon deactivation of the laser.	Photosensitizer solutions used in APDT systems, such as those containing methylene blue or toluidine blue, for disinfection for a medical purpose do not qualify as medical devices. This decision is based on the following: • their primary action is not physical; • their activation to provide an anti-bacterial function is not achieved via physical means; • their composition, -they are typically sold separately from the laser. Lasers used in APDT systems for disinfection for a medical purpose do qualify as medical devices. Such lasers are classified as active therapeutic devices under Rule 9 and are Class IIa, unless energy is transferred in a hazardous manner, in which case they are Class IIb. The laser (i.e., laser generator) and photosensitizer system, when used for a medical purpose, are not considered to be an integral product at the time of use and, therefore, cannot qualify as a Class III medical device under Rule 13.
Tissue expanders used in the breast	The product is intended by the manufacturer to be introduced under the skin in the breast area, during a breast reconstruction or breast augmentation when there is not enough skin to accommodate a permanent implant. The tissue expander consists of an expansion envelope with fill tube and an inflation valve. The device is filled every week or fortnight with saline solution causing the skin to stretch and grow until the achievement of the required volume. Once the skin has expanded sufficiently (generally in 6 to 9 months), the device is removed and a permanent breast implant is placed. The directive 2003/12/EC required that the breast implants should be classified as Class III medical devices. The question under consideration is since the breast expanders are introduced in the breast area should these devices be covered by Directive 2003/12/EC.	The breast expander is implanted into the body by surgical intervention and is intended to remain in place after the procedure for at least 30 days, the tissue expander is an implantable device. Considering that this implantable device is placed in the breast tissue, it should be considered that a breast tissue expander is a breast implant. Thus, by application of the Directive 2003/12/EC, the breast tissue expander should be classified as a Class III medical device.

Dura guard for use with a craniotome	The cranial dura mater is the outermost layer of the meninges that cover the brain. It lines the interior of the skull. The outer surface of the cranial dura mater adheres closely to the inner surface of the cranial bone. The dura guard device is intended to be attached to the craniotome during a craniotomy procedure to prevent the drill of the craniotome from damaging the dura mater.	The dura guard device is for use with a craniotome and is a device in its own right which is specifically intended by its manufacturer to protect the dura mater during a craniotomy procedure. To achieve its protective function, the device will come in contact with the inner surface of the cranial bone. The outer surface of the cranial dura mater is in contact with the inner surface of the cranial bone. As the dura guard will get into contact with the inner surface of the cranial bone, it will also come in direct contact with the cranial Dural mater. The contact with dura mater cannot be considered as accidental. Therefore, in normal conditions of use, the dura guard device is specifically for use in direct contact with the central nervous system (dura mater) and should be classified as a Class III medical device in accordance with Rule 6 of Annex IX of Medical Devices Directive.
Heart bypass cannula	Open heart surgery is performed with the use of the heart-lung machine, also called cardiopulmonary bypass machine that can interrupt the circulation by taking over the functions of the heart and lung while the heart is stopped. The blood is shunted thanks to different catheter/cannula in contact with the heart of the central circulatory system. The surgery is performed for a duration of under 6 hours. The aortic cannula is inserted in the aorta and is intended to carry oxygenated blood from the hemofilter to the heart. The venous cannula is inserted in the vena cava and/or the right atrium and is intended to drain non-oxygenated blood from the heart to the reservoir.	The aortic cannula and the venous cannula are specifically intended to be used in cardiopulmonary bypass surgery, in a context of a defect of the heart (stopped heart). They are intended to channel the blood from the heart of the central circulatory system and back into the central circulatory system in order to replace heart function. In order to establish circulation, a cardiopulmonary bypass circuit is created, and the venous and aortic cannula are parts of the essential elements of the circuit. It is considered that maintenance of blood circulation using a circuit whose parts are in direct contact with the central circulatory system and/or the heart constitutes control of a heart defect. Besides, the Medvedev 2.4/1 rev.9 quotes "thoracic catheters intended to drain the heart" as examples of Class III medical devices per Rule 7. On the basis of above, aortic perfusion cannula and venous drainage cannula should be classified as Class III devices per Rule 7 of Annex IX of Directive 93/42/EEC.

8.29. Liquid nitrogen for cryopreservation of cells and tissues of human origin for medical purpose	Liquid nitrogen, packaged in closed pressurized containers available in different sizes or in fixed closed pressurized tanks connected by a vacuum pipe to cryogenic containers, is used for preservation of cells and tissues of human origin, intended to be re-implanted or re-used in the human body. Liquid nitrogen has a medical purpose, even if direct contact with biological material does not occur.	Liquid nitrogen for cryopreservation of tissues and cells of human origin, intended to be re-implanted or re-used in the human body, should be regulated as a medical device although there is no direct contact with tissues and cells. For this intended use liquid nitrogen should be classified in Class IIa based on Rule 2.
Whole-body and partial body cryotherapy chambers	Cryotherapy chambers deliver intense cold (around –110°C) to the body for short periods of time (around 3 minutes). The whole body could be exposed (Whole Body Cryotherapy-WBC) or the body with the exception of the head (Partial Body Cryotherapy-PBC). Cryotherapy chambers which are medical devices are intended by their manufacturers to be used for pain relief, limitation of edema post-surgery, treatment of rheumatologic pathologies and muscular injuries, and/or for reduction of inflammation.	WBC and PBC devices generate a temperature around –110°C. The site of action of the WBC and PBC devices is the whole body, including the extremities such as feet, hands, and, for whole body cryotherapy, the head, which are particularly sensitive to the cold. According to contraindications given by the manufacturers, the use of such devices may affect the vital physiological functions of the patient. Their use entails potential hazards: asphyxiation, hypoxia, or oxygen deficiency, as well as frostbite, burns, and eye injury. As these devices exchange energy with respect to the human body in a potentially hazardous way, they should be classified as active therapeutic devices Class IIb according to Rule 9.
Trial hip prosthesis heads or stems	Trial hip prosthesis heads or stems are medical devices intended to be used during surgery to determine the exact size of the prosthesis before the implantation of the definitive implant. They are surgically invasive devices for transient use. The definition in the Directive 93/42/EEC on medical devices of "reusable surgical instrument" is "instrument intended for surgical use by cutting, drilling, sawing, scratching, scraping, clamping, retracting, clipping or similar procedures […]."	Evaluating the appropriate size of a prosthesis prior to implantation cannot be considered as a "similar procedure" to those mentioned in the definition of surgical instruments. Consequently, trial prostheses do not fit the definition of reusable surgical instruments. Thus, trial prosthesis heads and stems are surgically invasive devices but not reusable surgical instruments. Consequently, they should be classified as Class IIa devices as per Rule 6.

Lubricants intended for alleviation of vaginal dryness	Water- or silicone-based lubricants intended by their manufacturer for the alleviation of vaginal dryness during sexual intercourse should be qualified as medical devices since they are intended for replacement of a physiological process and thereby comply with the definition of medical device.	After sexual intercourse, it is expected, or at least likely, that part of the lubricant will remain in the body for more than 60 minutes and not more than 30 days. This is an inherent characteristic of the product and is considered part of its intended (normal) use and not as accidental use (as stated in Section 3.2 of MEDDEV 2.4/1 rev. 9). Lubricants intended for the alleviation of vaginal dryness during sexual intercourse are therefore invasive devices intended for short term use (more than 60 minutes and not more than 30 days), unless demonstrated otherwise by the manufacturer. According to Rule 5, these lubricants should be classified in Class IIa. In case the manufacturer can provide sufficient evidence to prove that the lubricant does not remain in the human body beyond 60 minutes after application, the lubricant may be considered to be intended for transient use and should then be classified in Class I (Rule 5).

10.17. SOFTWARE AND MOBILE APPLICATIONS

Software and mobile applications that fall under the definition of a medical device or an *in-vitro* diagnostic medical device are regulated by the respective Directives 93/42/EEC or 98/79/EC. There is a MEDDEV Guidance 2.1/6, entitled "Guidelines on the qualification and classification of stand-alone software used in healthcare within the Regulatory Framework of Medical Devices" that seems to be the most relevant document which provides practical advice to manufacturers and other economic operators on how to determine when a software falls under the definition of a medical device or of an *in-vitro* diagnostic medical device. These criteria also apply to mobile applications.

A mobile application for processing ECGs	**This application -or app- (software program) allows clinicians to review a patient's ECG history during routine check-ups and see a full presentation generated by the analysis algorithms. ECG-related and supplementary data can be received wirelessly, for example from the hospital ECG management system.** **The app's intended use is to allow for more timely and accurate diagnosis and treatment of the patient.**	**This app, which is not incorporated into a medical device, uses signal data from an external source and processes it to an ECG waveform, thereby performing an action on data other than just storage, for the medical benefit of individual patients.** **In accordance with the guidance document MEDDEV 2.1/6 rev. 113, such a mobile app for viewing ECGs should be qualified as standalone Class IIa medical device, according to Rule 10 of Annex IX to Directive 93/42/EEC.**
A mobile application for the communication between patient and caregivers while giving birth	**This app allows storing data on each moment of contraction during delivery, after a healthcare professional established that a woman is in the process of giving birth. Users can make notes and take pictures for the purpose of exporting them to an external website also providing documentation for a later stage.** **The app's intended use is to improve the quality of communication between the patient and caregivers.**	**This app, which is not incorporated in a medical device, performs an action on data limited to storage and simple search.** **In accordance with the guidance document MEDDEV 2.1/6 rev.1, a mobile app for the communication between patients and caregivers while giving birth should not be qualified as a medical device.**
A mobile medical application for viewing the anatomy of the human body	**This app allows users to view illustrations of the anatomy of the human body. In addition, it introduces the reader to anatomical terminology, positioning, and medical imaging.**	**Although the app is not only performing simple data search, it is not used directly for the medical benefit of the individual patients. In light of this, this app should not be qualified as a medical device.**

Qualification of software for interpretation of a guideline	**The product in question consists of a software application which allows for faster consulting/reading of an international guideline regarding the Classification of Malignant Tumors (TNM) issued by the UICC (International Union Against Cancer), in the view of classification of cancer by anatomic disease extent.** **Three variables are introduced into the system:** **•** **the size of the primary tumor and whether it has invaded nearby tissue (described as T0, T1, T2, etc.);** **•** **regional lymph nodes that are involved (indicated as N0, N1, N2, etc.);** **•** **metastasis (M0 or M1).** **According to the selection of these 3 categories/variables, the application indicates the disease's stage of development (cancer extent), according to the above-mentioned guideline.** **According to the manufacturer, this information software simply facilitates the search and use of an international guideline which physicians usually consult via electronic file, or in paper format.**	**The software does not perform an action on data other than simple search function, as per in MEDDEV Guidance 2.1/6.** **The product does not therefore fulfill the definition of medical device, according to Directive 93/42/EEC, and should not be qualified as such.**
Qualification and classification of software for delivery and management of cognitive remediation and rehabilitation programs	**This software is designed to deliver and manage cognitive remediation and rehabilitation programs and is intended to be used for the treatment of a variety of neurotrauma, neurodegenerative, and neuropsychiatric conditions.** **The software contains different programs designed for targeted stimulation of cognitive functions, by using interactive games and exercises. Such cognitive functions include executive functions (reasoning and strategy), verbal memory, visual memory, spatial memory, auditory memory, processing speed. This software also allows the clinician to plan, monitor, and assess the patient's progress throughout the treatment plan.**	**Based on the intended purposes, namely the treatment of disease, injury or handicap, this software should be qualified as a medical device. When such software qualifies as a medical device, it shall be classified as Class I medical device according to Rule 12 of Directive 93/42/EEC.**

Classification of software for information management and patient monitoring	**The product in question is a software-based system that includes an informatics management and patient monitoring platform. The informatics management component of the software is intended to route and store medical device data and device diagnostic information from supported bedside devices to the electronic medical record and clinical information system. The patient monitoring platform is intended to allow health care professionals remotely consult a patient's status and remotely review other standard or critical near real-time patient data, waveforms, and alarms in order to utilize this information to support clinical decisions and deliver patient care in a timely manner.** **The patient monitoring platform has the following functions:** **1.** **View collated patient information and track changes in patient history;** **2.** **Generate audible alerts when an alarm is coming through the system;** **3.** **Generate trend graphs which include red-shading when an alarm is sounding;** **4.** **Set patient-specific "filtering rules" based on alarm severity and alarm type. These alarm rules support the ability to suppress specific alarms, delay specific alarms and separate rules for visual, audio, and paging of alarms;** **5.** **Send SMS and e-mails and filter alarm forwarding to 3rd party annunciation systems.**	**This system is intended to be used in intensive care wards with ventilators, pulse oximeters, and other devices used for monitoring patients, although similar systems could be used with devices for monitoring non-vital physiological processes. Per MEDDEV 2.12/6 clinical information systems are not qualified as medical devices. The patient monitoring platform contains a number of different functions, each of which must be assessed when determining whether the software meets the definition of a medical device using the criteria described in MEDDEV 2.1/6.** **Per MEDDEV 2.1/6 "if the software does not perform an action on data, or performs an action limited to storage, archival, communication, "simple search" or lossless compression (i.e., using a compression procedure that allows the exact reconstruction of the original data) it is not a medical device." This is applicable for the listed functions performed by the patient monitoring platform with the exception of the alarm filtering function. The software is not considered to be "generating" an alarm as it is the bedside device that generates the alarm based on its analysis of patient physiological data. The bedside device also assigns a severity to that alarm. While this software does not generate the original patient alarm, it applies user-defined filtering rules to each alarm category (e.g., severity) received by the software. This filter function is considered to be performing a search of the nearly "live" data received from the bedside device that results in a specific action being taken on that alarm, i.e., alarm is delayed. The action of the filter function is not considered a "simple search" of archived data. The delay to the alarm that results from the filter function is considered to lead to the generation of new or additional information that contributes to the monitoring and follow-up of the patient connected to the bedside device. If an alarm is noted on the system, the users are instructed to interact with the bedside device, which would be considered to be influencing the use of the bedside device.** **Therefore, this software, having one of its functions qualified as a medical device, complies with the definition of a medical device and should be qualified as such. When classifying this device implementing Rule 2.3 of Annex IX applies.**

Mobile application for managing pictures of moles	This app allows users to take pictures of moles on the skin and subsequently store them for the purpose of managing these pictures and showing them to a doctor when it is suspected that visible characteristics of the mole change over time.	This app, which is not incorporated in a medical device, does not perform an action on data other than just storage. In accordance with the guidance document MEDDEV 2.1/6 rev. 1, this mobile app should not be qualified as standalone medical device software.
Mobile application for the assessment of moles	This app allows users to take pictures of moles on the skin and subsequently store and compare them. Based on image processing algorithms, the app provides a detailed assessment of the scanned moles. Additionally, the app assesses the probability that the scanned mole is a melanoma in order to support early diagnose of skin cancer.	This app, which is not incorporated in a medical device, uses computer image processing technology to make assessments of the moles, whereby performing an action on data other than just storage, for the medical benefit of individual patients. In accordance with the guidance document MEDDEV 2.1/6 rev. 1, a mobile app for the assessment of moles should be qualified as standalone Class I medical device, according to Rule 12 of Annex IX to Directive 93/42/EEC.
Product intended to facilitate conception based on basal body temperature	The product in question measures the basal body temperature (BBT) orally, records it and uses the daily BBT and menstruation days to track the menstrual cycle and predict ovulation. Once it has enough data from the user, it calculates the user's fertility status based on a validated statistical algorithm as claimed by the manufacturer. The fertility status of the current day is reflected by one of three indicator lights: red (fertile), green (infertile) or yellow (learning phase/cycle fluctuation). The temperature sensor, the activation button (interface), the processor (storing of data and fertility status calculator) and the menstruation and fertility status indicators are all integrated into a single piece of equipment. The product is battery-driven. It does not display the user's temperature and is not intended to allow direct diagnosis or monitoring of vital physiological processes, i.e., it is not intended to be used as an electronic thermometer. It is claimed by the manufacturer to facilitate conception by predicting ovulation.	The product is intended by the manufacturer to facilitate conception and should be qualified as a medical device. According to clauses 1.2 and 1.4 of Chapter I, Annex IX MDD, this device is considered to be an invasive active medical device and should be classified as Class I according to Rule 12.

Product intended to facilitate conception and enable contraception based on basal body temperature	**The product in question measures the basal body temperature (BBT) orally, records it and uses the daily BBT and menstruation days to track the menstrual cycle and predict ovulation. Once it has enough data from the user, it calculates the user's fertility status based on a validated statistical algorithm as claimed by the manufacturer. The fertility status of the current day is reflected by one of the three indicator lights: red (fertile), green (infertile) or yellow (learning phase/cycle fluctuation).** **The temperature sensor, the activation button (interface), the processor (storing of data and fertility status calculator) and the menstruation and fertility status indicators are all integrated into a single piece of equipment. The product is battery-driven. It does not display the user's temperature and is not intended to allow direct diagnosis or monitoring of vital physiological processes, i.e., it is not intended to be used as an electronic thermometer.** **The product in question is claimed by the manufacturer to facilitate conception and prevent pregnancy.**	**The product is intended by the manufacturer to (i) facilitate conception or (ii) prevent pregnancy, and should therefore be qualified as a medical device.** **According to clause 4.2 of Chapter III, Annex IX MDD, all devices used for contraception or the prevention of the transmission of sexually transmitted diseases are in Class IIb (Rule 14). The explanations of Rule 14 in the MEDDEV 2.4/1 Rev. 9 emphasize that "the intended uses relate to special cases of human vulnerability that cannot be covered by the normal criteria of time, invasiveness, and organic function." Unwanted pregnancy is a case of human vulnerability.** **According to clauses 1.2 and 1.4 of Chapter I, clause 2.5 of Chapter II and clause 4.2 of Chapter III, Annex IX MDD, and in consideration of the above-mentioned rationale, this device is considered to be an invasive active medical device and should be classified as Class IIb according to Rule 14.**
Stand-alone software application for conception and contraception purposes using data entered by the patient	**The product in question is claimed by its manufacturer to be a natural method of birth control. It is a stand-alone software application that combines the calendar/rhythm method, the body basal temperature (BBT) method and the cervical mucus method to both (i) prevent pregnancy and (ii) target the most fertile time for getting pregnant. The user enters her data (first day of menstruation, BBT measured with a common thermometer and consistency of the cervical mucus) and obtains the fertility window.**	**The product is intended by the manufacturer to (i) prevent pregnancy or (ii) facilitate conception, and should be qualified as a medical device.** **According to clause 1.4 of Chapter I, clause 2.5 of Chapter II and clause 4.2 of Chapter III, Annex IX MDD, this stand-alone software is considered to be an active medical device and should be classified as Class IIb according to Rule 14.**

Medication decision support software	**The product is software intended to be used by healthcare professionals for optimizing patient medicinal therapy by identifying possible contraindications, medicinal product interactions, and the need for dosage adjustments.** **The software uses a patient's clinical data and applies a number of rules, of varying complexity, to generate an output. The simplest rules may depend on a single condition being met. These typically use data relating to prescribed medication that is stored in the internal dataset. Other rule-based principles may be more complex and may depend on several conditions being met for a recommendation to be made. Certain rules can have preference for one condition over another.** **The output information from the software is used by the healthcare professional to refine and optimize the individual patient's medicinal therapy treatment plan, e.g., to give advice to remove a particular medicinal product to avoid possible interactions, to highlight adverse reactions to medicinal products, to provide treatment options for previously untreated conditions, etc.**	**The software is intended to be used for the purpose of prevention, monitoring, treatment, or alleviation of a disease and should therefore be qualified as a medical device.**

10.18. CONCLUSION

With the focus of many regulatory authorities often falling on medicines rather than medical devices, the need for finding ways to improve the quality and access to safe medical devices has increased. For many countries, their regulatory systems have proven to be ineffective in quality management and in monitoring the domestic market so as to prevent unsafe and low-quality medical products from entering. The establishment of an effective regulatory system is dependent on numerous factors including long term government commitment and an overall political and societal willingness to cooperate. However, with that being said, one of the first steps to ensuring safe and effective devices are placed on the market starts with the manufacturer understanding the overall intended use of their device. This subsequently allows the manufacturer to classify the device correctly based on the risk it brings. When exporting to Europe, manufacturers must therefore be aware of, and ensure they understand the requirements of the European Regulations as well as related guidance documents in order to select the right conformity assessment route and demonstrate full compliance.

Bibliography

1. American Medical Association, (2003) World medical association declaration of Helsinki. ethical principles for medical research involving human subjects. *Clinical Review and Education*, 2191–2194.
2. *Application of Risk Management to Medical Devices,* (2007). International Organization for Standardization (ISO 14971:2007 Standard). https://www.iso.org/obp/ui/#iso:std:iso:14971:ed-2:v2:en (accessed on 7 June 2021).
3. Bloom, G., & Temple-Bird, C., (1990). *Medical Equipment in Sub-Saharan Africa: A Framework for Policy-Formulation.* IDS Research Report No. 19, WHO/SHS/NHP/90.6, WHO, Geneva.
4. Carlson R V., Boyd, K. M., & Webb, D. J., (2004). The revision of the declaration of Helsinki: Past, present and future. *British Journal of Clinical Pharmacology, 57*, 695–713.
5. Clinical Investigations, (2010). *Global Harmonization Task Force (GHTF/SG5/ N3:2010).* http://www.imdrf.org/docs/ghtf/final/sg5/technical-docs/ghtf-sg5-n3-clinical-investigations-100212.pdf (accessed on 7 June 2021).
6. Commission Implementing Decision M/565 of 15.5.2020 on a standardization request to the European Committee for Standardization and the European Committee for Electrotechnical Standardization as regards medical devices in support of Regulation (EU) 2017/745 of the European Parliament and of the Council and *in vitro* diagnostic medical devices in support of Regulation (EU) 2017/746 of the European Parliament and of the Council.
7. *Development of Medical Device Policies,* (2011). Geneva: World Health Organization. WHO Medical Device Technical Series. http://apps.who.int/medicinedocs/documents/s21559en/s21559en.pdf (accessed on 7 June 2021).

8. EU Council Directive 93/42/EEC, (1993). *Concerning Medical Devices*. http://www.imdrf.org/docs/ghtf/archived/sg4/technical-docs/ghtf-sg4-guidelinesauditing-qms-part-1-general-requirements-080827.pdf (accessed on 7 June 2021).
9. Falcitelli, G., (1994). Biomedical technology and clinical engineering in cooperation programs with developing countries: Methodological guidelines. In: *Proceedings of the Regional Workshop on Health Care Technology in the Sub-Saharan Region* (p. 13). Somerset West, South Africa. April 1994 South African Medical Research Council, Tygerberg.
10. Ghtf.org. *Global Harmonization Task Force*. Generating the GMDN Process and Development. Available from: http//www.ghtf.org/meetings/conferences/8thconference/.../Dillard.ppt (accessed on 7 June 2021).
11. *Global Medical Devices Nomenclature System (GMDN)*. [website] https://www.gmdnagency.org (accessed on 7 June 2021).
12. *Good Governance for Medicines. Model Framework*, (2014). Geneva: World Health Organization. http://apps.who.int/iris/bitstream/10665/129495/1/9789241507516_eng.pdf?ua=1&ua=1 (accessed on 7 June 2021).
13. *Guiding Principles for Regulatory Quality and Performance,* (2005). Paris: Organization for Economic Cooperation and Development (OECD). http://www.oecd.org/fr/reformereg/34976533.pdf (accessed on 7 June 2021).
14. Issakov, A., (1994). Service and maintenance in developing countries. In: Van, G. C.W. D. (ed.), *Medical Devices: International Perspectives on Health and Safety* (pp. 21–28). Elsevier, Amsterdam.
15. Loh, & Boumans, (2018). *Understanding Europe's New Medical Devices Regulation (MDR 2017/745): New Requirements, Key Changes and Transition Strategies for Device Companies*. Emergo White Paper.
16. *Manual on Borderline and Classification in the Community Regulatory Framework for Medical Devices,* (2019). https://ec.europa.eu/health/sites/health/files/md_topics-interest/docs/md_borderline_manual_05_2019_en.pdf (accessed on 7 June 2021).
17. *MEDDEV 2.4/1 Rev 9 Classification*, (2020). https://www.cepartner4u.com/download/meddev-2-41-rev-9-classification/ (accessed on 7 June 2021).

18. *Medical Device Regulations: Global Overview and Guiding Principles,* (2003). Geneva: World Health Organization. http://www.who.int/medical_devices/publications/en/MD_Regulations.pdf (accessed on 7 June 2021).

19. MHRA (Medicines and Healthcare products Regulatory Agency), (2010a). *Medical Device Alert: DePuy ASR Acetabular Cups Used in Hip Resurfacing Arthroplasty and Total Hip Replacement.* MDA/2010/044. 25 May 2010.

20. *Official Journal of the European Union: Commission Communication in the Framework of the Implementation of the Council Directive 93/42/EEC Concerning medical Devices.* (Publication of titles and references of harmonised standards under Union harmonisation legislation) Text with EEA relevance: (2017/C 389/03).

21. Official Journal of the European Union: Commission Implementing Decision (EU) 2020/437 of 24 March 2020 on the harmonized standards for medical devices drafted in support of Council Directive 93/42/EEC.

22. *Principles of Conformity Assessment for Medical Devices,* (2012). Global harmonization task force, (GHTF/SG1/N78:2012). http://www.imdrf.org/docs/ghtf/final/sg1/technical-docs/ghtf-sg1-n78-2012-conformity-assessment-medicaldevices-121102.pdf (accessed on 7 June 2021).

23. *Principles of Medical Devices Classification,* (2012). Global harmonization task force; (GHTF/SG1/N77:2012). http://www.imdrf.org/docs/ghtf/final/sg1/technicaldocs/ghtf-sg1-n77-2012-principles-medical-devices-classification-121102.pdf (accessed on 7 June 2021).

24. *Quality Management System-Medical Devices-Guidance on Corrective Action and Preventive Action and Related QMS Processes* (2010). Global harmonization task force; (GHTF/SG3/N18:2010). http://www.imdrf.org/docs/ghtf/final/sg3/technical-docs/ghtf-sg3-n18-2010-qms-guidance-on-corrective-preventativeaction-101104.pdf (accessed on 7 June 2021).

25. Regulation (EU) 2017/745 of the European Parliament and of the Council of 5 April 2017 on medical devices, amending Directive 2001/83/EC, Regulation (EC) No 178/2002 and Regulation (EC) No 1223/2009 and repealing Council Directives 90/385/EEC and 93/42/EEC.

26. *Role of Standards in the Assessment of Medical Devices,* (2008). Global harmonization task force; (GHTF/SG1/N044:2008). http://www.imdrf.org/docs/ghtf/final/sg1/procedural-docs/ghtf-sg1-n044-2008-standards-in-assessment-of-medicaldevices-080305.pdf (accessed on 7 June 2021).
27. *Safety Aspects - Guidelines for Their Inclusion in Standards,* (2014). Geneva: International Organization for Standardization; (ISO/IEC Guide 51:2014. http://www.iso.org/iso/catalogue_detail.htm?csnumber=53940 (accessed on 7 June 2021).
28. Temple-Bird, C. *Methodology for the Analysis of a National Medical Equipment Management System Based on a Case Study in Botswana* (p. 33). ibid.
29. WHO. *Global Governance.* http://www.who.int/healthpromotion/conferences/9gchp/good-governance/en/ (accessed on 7 June 2021).
30. WHO. *Good Regulatory Practices: Guidelines for National Regulatory Authorities for Medical Products*. Geneva: World Health Organization.
31. World Health Organization, (2001). *Bulletin of the World Health Organization*. World Medical Association Declaration of Helsinki. Ethical Principles for Medical Research Involving Human Subjects.

Index

D

E

F

G

H

I

L

M